Health Informatics

This series is directed to healthcare professionals leading the transformation of healthcare by using information and knowledge. For over 20 years, Health Informatics has offered a broad range of titles: some address specific professions such as nursing, medicine, and health administration; others cover special areas of practice such as trauma and radiology; still other books in the series focus on interdisciplinary issues, such as the computer based patient record, electronic health records, and networked healthcare systems. Editors and authors, eminent experts in their fields, offer their accounts of innovations in health informatics. Increasingly, these accounts go beyond hardware and software to address the role of information in influencing the transformation of healthcare delivery systems around the world. The series also increasingly focuses on the users of the information and systems: the organizational, behavioral, and societal changes that accompany the diffusion of information technology in health services environments.

Developments in healthcare delivery are constant; in recent years, bioinformatics has emerged as a new field in health informatics to support emerging and ongoing developments in molecular biology. At the same time, further evolution of the field of health informatics is reflected in the introduction of concepts at the macro or health systems delivery level with major national initiatives related to electronic health records (EHR), data standards, and public health informatics.

These changes will continue to shape health services in the twenty-first century. By making full and creative use of the technology to tame data and to transform information, Health Informatics will foster the development and use of new knowledge in healthcare.

More information about this series at http://www.springer.com/series/1114

Adrian Stavert-Dobson

Health Information Systems

Managing Clinical Risk

Adrian Stavert-Dobson
Sheffield
UK

ISSN 1431-1917 ISSN 2197-3741 (electronic)
Health Informatics
ISBN 978-3-319-26610-7 ISBN 978-3-319-26612-1 (eBook)
DOI 10.1007/978-3-319-26612-1

Library of Congress Control Number: 2015959797

Springer Cham Heidelberg New York Dordrecht London

Printed on acid-free paper

Springer International Publishing AG Switzerland is part of Springer Science+Business Media (www.springer.com)

For Alice and Thea

Foreword

Health information technology is an integral part of our healthcare system and is invaluable to the safety-critical nature of the NHS. Over the years, health IT has become increasingly central to how we effectively deliver care – and in empowering our patients to take an interest in and learn about their own health quickly and conveniently.

As a GP, I have seen how technology has transformed the way we deliver care and supported us to practise more safely, for example by implementing electronic medical records, and through electronic prescribing, which by simply eliminating illegibility in drug names and doses on handwritten prescriptions, has allowed us to avoid a large proportion of medication errors. There is the potential for so much more.

New technology can also help by reducing burdens on healthcare professionals, right across the health service by making processes as straightforward as possible, and encouraging patients to engage in 'self-care', where appropriate. These are positive developments, but patient safety must always come first.

As well as all the potential benefits of new technology in healthcare, health IT also has the potential to introduce new errors that were not previously possible, for example, selecting the wrong option from drop-down lists when completing electronic forms.

It is therefore essential to develop comprehensive safeguards around any new technological developments to establish a strong safety culture and ensure that our patients always receive the best possible, safe care. These responsibilities fall both to IT suppliers and to the healthcare organisations that implement the solutions.

We cannot turn away from the inevitable future of medicine – with endless possibilities for health services and our patients, I don't think we would want to.

The NHS is evolving and it is important that healthcare professionals across the health service evolve with it. This book provides a helpful and timely introduction to the principles and practice of developing inherently safe systems for healthcare.

London, UK

Professor Maureen Baker, CBE, DM, FRCGP, DRCOG, DCH
Chair of Council, Royal College of General Practitioners
Strategic Safety Adviser
Health and Social Care Information Centre

Preface

Life as a junior doctor in 1997 was, for me at least, bittersweet. My life-long dream of working in the UK's National Health Service caring for individuals often at their most desperate time of need was both humbling and an honour. But something about the way we delivered healthcare irked me, something which over time came to eat away at my very soul. And that was the unimaginably inefficient and archaic processes we employed every day to deliver care. I imagined the wellbeing of those I cared for to be at the centre of my day: informing, reassuring, comforting. But this vision was an unsuspecting naïveté.

In reality life as a junior doctor was filled with endless administration. At times it felt like my very being was defined not by my aptitude to care but by my ability to manage information. My productivity and effectiveness, measured in terms of anything which really mattered to my patients, was frankly minimal.

At every stage of the labour intensive and uncoordinated processes was the opportunity for human error. Like all junior doctors I made mistakes, mistakes which I so desperately wanted to learn from. But I operated in an environment which did little or nothing to take account of human error. There simply had to be a better way and what was particularly frustrating for me was that I could see it, almost touch it.

As a teenager, computer programming had been my life. To me the logic, control and complexity of computer code held a strange beauty and elegance. For over a decade I had authored software in education, music, science and pharmacy – a grounding in commercial software development which also happened to offer a curious hedonism only experienced by fervent programmers.

With technology I knew I had an opportunity to make a difference on a grander scale, changing the lives not of individual patients but of entire populations. I embraced Health IT as a career and when I saw what it could achieve I never looked back. By handing over the information layer to technology I could free up valuable time for my colleagues. I could get data to care providers and their patients at the right time without the dependency on manual transportation and archaic paging systems. I could offer rich electronic decision support to overcome the limitations of any individual human's intellect. And, most of all, I could facilitate care being delivered safely.

But one bubble was eventually to burst. Around 2004 it was becoming increasingly apparent in this rapidly evolving industry that Health IT systems could bring problems of their own. For those software products outside the formalities of medical device regulation, there was little or no clinical risk management in place and yet Health IT had the potential to cause real harm to individuals. I witnessed electronic x-ray studies labelled with the wrong patient's name, blood test results with misleading values and prescriptions with confusing drug dosages. The causes of these issues were utterly intriguing – not simple bugs as one might imagine but a complex relationship between requirements, design, configuration, training, documentation and change management. Once again I was hooked and I soon came to realise that unless this complex web of system and human failure could be consistently characterised and untangled, the role of IT in healthcare had very definite constraints.

Around 2006, I was lucky to work with a number of organisations who prioritised safety in Health IT and had the pleasure of meeting talented safety engineers who taught me to understand risk. I came to realise that the hazards which Health IT introduced did not impose limits on the technology but simply represented more fascinating challenges thoroughly solvable by human intellect. What's more, in the pursuit of risk reduction one invariably creates a product of higher quality, reliability and ultimately market value. When risk is properly tackled it allows Health IT to flourish. Only then can users realise the technologies' true benefits whilst simultaneously protecting those for whom we care.

I wrote this book to fill a gap. There is much in the literature about patient safety, medical informatics and the engineering of safety critical systems, but outside of medical device regulation little is said about how we manufacture and implement Health IT safely. In this book I have attempted to consolidate what the industry has learnt over a 10–15-year period. As technologies and techniques evolve it is clear that this is just the start of an exciting journey, the birth of an academic discipline which brings together learnings from many disparate sources.

In Health IT we have an enormously powerful tool – one with overwhelming potential but which has to be handled with the care and respect it deserves. Through a thorough understanding of clinical medicine, information technology and engineering we have it in our power to take safe innovation in healthcare to the next level for future generations.

I would like to thank all those who have contributed to and supported me in creating this publication – in particular Dr. Chris Vincent, Prof. George Despotou, Prof. Maureen Baker, James Savage and Peggy Edwards. Finally I offer eternal gratitude to my family especially my wife Jo for her unwavering support during the 3 years it has taken to bring this book to fruition.

Sheffield, UK Adrian Stavert-Dobson

Contents

Part I Risk and Safety

1 Introduction 3
1.1 What Is Patient Safety? 3
1.2 The History of Patient Safety 4
1.3 Health IT 6
1.4 HIT Impact on Patient Safety 7
1.5 Controlling Risks from HIT 12
1.6 Regulation of Health IT Systems 14
1.6.1 Business Management Systems 14
1.6.2 Care Administration Systems 15
1.6.3 Clinical Management Systems 16
1.6.4 Medical Devices 16
1.6.5 Health Applications Aimed at Mobile Devices 18
1.7 Summary 19
References 19

2 Risk and Risk Management 23
2.1 Overview of Risk 23
2.2 Types of Risk 24
2.3 The Perception of Risk 25
2.4 Perception of Risk in HIT 26
2.5 Risk Management 27
2.6 Hazards 28
2.7 Components of Risk 29
2.7.1 Risk and Probability 30
2.8 Quantifying Risk 31
2.8.1 Developing Risk Matrix Categories 32
2.8.2 Severity 34
2.8.3 Likelihood 35
2.8.4 Risk Equivalence 37

2.9 Summary 37
References 37

3 **Acceptability and Ownership of Risk** 39
3.1 How Safe Is Safe? 39
3.2 Acceptability Stratification 41
3.3 Risk Acceptability of Medical Devices 42
3.4 Who Owns the Risk? 43
3.5 Risk Versus Benefits 44
3.6 Summary 45
References 46

4 **Standards and Quality Management in Health IT** 47
4.1 What Are Standards? 47
4.2 Examples of Risk and Quality Management Standards 48
4.2.1 IEC 61508 48
4.2.2 ISB 0129 and ISB 0160 (UK) 49
4.2.3 ISO/TS 25238:2007 51
4.2.4 ISO 14971:2012 52
4.2.5 IEC 80001-1:2010 52
4.2.6 ISO 9000 and ISO 13485 53
4.2.7 ISO 62366:2007 55
4.2.8 IEC 62304:2006 56
4.3 Summary 57
References 57

Part II The Causes of Risk in Health IT

5 **Safety and the Operating Environment** 61
5.1 Health IT and the Socio-technical Environment 61
5.2 Building a Safety Culture 65
5.3 Human Failure 66
5.3.1 Human Error 66
5.3.2 Violation 68
5.4 Human Error and User Interface Design 69
5.4.1 Design Methodology 70
5.4.2 Minimising Errors by Design 71
5.4.3 Mental Models and Predictive Behaviour 75
5.5 Incident Reporting 76
5.6 Summary 77
References 78

6 **Failure of Health IT** 81
6.1 Dependable Computing 81
6.2 System Failure 82

6.3 Causes and Failure Modes ... 82
6.4 Classification of Causes ... 83
6.5 Absent Clinical Data ... 84
6.5.1 Complete Service Failure ... 85
6.5.2 Module and Functional Failure ... 85
6.5.3 Data Delivery Failure ... 86
6.5.4 Poor System Performance ... 86
6.5.5 Inappropriate Denial of Access ... 86
6.5.6 Elusive Data ... 87
6.6 Misleading Clinical Data ... 88
6.6.1 Misleading Identity ... 89
6.6.2 Misleading Context ... 90
6.6.3 Misleading Presentation ... 91
6.6.4 Misleading Input ... 94
6.6.5 Data Corruption ... 95
6.6.6 Miscommunication ... 97
6.6.7 Misleading Direction ... 98
6.7 Summary ... 99
References ... 100

7 Availability and Performance of Health IT ... 101
7.1 Planned Service Outages ... 102
7.2 Unplanned Service Outages ... 102
7.3 Component Failure ... 103
7.3.1 Redundancy ... 104
7.3.2 Diversity ... 106
7.3.3 Service Monitoring ... 106
7.4 Catastrophic Failure scenarios ... 107
7.4.1 Distributed Architecture ... 107
7.4.2 Disaster Recovery ... 108
7.5 The Poorly Performing System ... 109
7.5.1 Time Wasting ... 110
7.5.2 System Abandonment ... 110
7.6 Avoiding and Managing Slow Performance ... 111
7.6.1 Factors Affecting Transactional Demand ... 111
7.6.2 Factors Affecting Processing Power and Bandwidth ... 113
7.7 Business Continuity Planning for Healthcare Organisations ... 114
7.7.1 Unavailability of Existing Clinical Data ... 115
7.7.2 Capture of New Clinical Data ... 115
7.7.3 Communication of Critical Clinical Data ... 116
7.8 Summary ... 116
References ... 117

Part III Getting Your Organisation Ready

8 The Safety Management System 121
8.1 What Is a Safety Management System? 121
8.2 Types of Safety Management Systems 123
8.3 Documenting the Safety Management System 125
8.4 The Clinical Risk Management Policy 126
8.5 The Safety Management System Processes 128
8.6 Governing the Safety Management System 129
8.6.1 Establishing Accountability 129
8.6.2 The Safety Management Committee 130
8.7 Auditing and Evaluating the Safety Management System 131
8.7.1 Auditing the SMS 131
8.7.2 Evaluating Effectiveness 133
8.8 Managing Process Variability 135
8.9 Document Management 137
8.10 Clinical Risk Management Training 137
8.11 Summary 138
References 138

9 The Scope of Clinical Risk Management 141
9.1 Information Governance (IG) Violation 142
9.2 Security Violation 142
9.3 Operation Outside of the Intended Purpose 143
9.4 Misuse 144
9.5 Harm to Individuals Other than Patients 144
9.6 Harm Arising from Patients Viewing Their Own Clinical Data . . 145
9.7 Assessing Absent Functionality 146
9.8 Medico-legal Issues 147
9.9 Summary 148
References 148

10 Evidencing a Competent Team 149
10.1 The Need for Competency 149
10.2 Competency Assessment Methods 150
10.2.1 Curriculum Vitae Repository 150
10.2.2 Skill Mapping 151
10.3 Competency and Resource Management 152
10.4 Summary 152
References 153

Part IV Undertaking a Clinical Risk Management Project

11 Planning a Clinical Risk Management Project 157
11.1 The CRM Plan 158
11.1.1 Defining the Project Boundaries 159
11.1.2 Intended Purpose 160

11.1.3 System Dependencies 161
11.1.4 "Black Box" Analyses 162
11.1.5 Retrospective Analyses 162
11.1.6 Managing Assumptions and Constraints 164
11.1.7 CRM in Projects with an Agile approach 165
11.1.8 Managing Project Change During Development 167
11.2 The Hazard Register 168
11.3 The Safety Case 169
11.3.1 What Is a Safety Case? 169
11.3.2 Benefits of the Safety Case 171
11.3.3 Potential Shortcomings of the Safety Case 172
11.4 Summary 173
References 173

12 Structuring the Hazard Register 175
12.1 Impacts of Hazards 175
12.2 Causes of Hazards 176
12.3 Controls 177
12.4 Bringing the Structure Together 179
12.5 Summary 181

13 Populating the Hazard Register 183
13.1 The Use of Formal Methods 183
13.2 Structured What-If Technique 184
13.2.1 Using Guidewords for Creative Thinking 184
13.2.2 The Multidisciplinary Approach 185
13.2.3 A Top-Down Methodology 186
13.3 Breaking Down the Scope 188
13.3.1 The Clinical Business Process 188
13.3.2 The System Business Process 189
13.3.3 The System Architecture 190
13.3.4 Interfaces and Messaging 191
13.3.5 Configuration and System Administrator Tasks 191
13.4 Brainstorming Ideas for Hazard Register Content 192
13.5 Using Storytelling in Hazard Derivation 195
13.6 Other Systematic Methods of Hazard Assessment 197
13.6.1 Failure Mode and Effects Analysis (FMEA) 197
13.6.2 HAZOP 199
13.6.3 Fault Tree Analysis and Event Tree Analysis 199
13.7 Safety-Related Issues 201
13.8 Summary 203
References 203

14 Estimating and Evaluating Clinical Risk 205
14.1 Factors Which Influence the Degree of Clinical Risk 205
14.1.1 Clinical Dependency 205
14.1.2 Workarounds 206

14.1.3 Detectability 208
14.1.4 Exposure and the Population at Risk 211
14.1.5 Complexity 212
14.1.6 Novelty 214
14.2 Bias in Risk Evaluation 215
14.3 Summary 217
References 218

15 Developing Control Strategies 219
15.1 Classification of Controls 220
15.1.1 Elimination 220
15.1.2 Substitution 221
15.1.3 Engineered Controls 221
15.1.4 Administrative Controls 222
15.1.5 Personal Protective Equipment? 222
15.2 Safety by Design 223
15.2.1 Active Engineered Controls 223
15.2.2 Passive Engineered Controls 224
15.3 Training and Human Factors as Controls 225
15.3.1 Identifying Training Controls 226
15.3.2 Classroom-Based Learning 226
15.3.3 Other Learning Modalities 230
15.4 Summary 231
References 232

16 Software Testing in Clinical Risk Management 233
16.1 Introduction 233
16.2 The Testing Process 234
16.2.1 Types of Software Testing 234
16.2.2 Phases of Software Testing 236
16.2.3 The Test Environment and Test Data 238
16.2.4 Testing Documentation 239
16.3 Testing as Evidence for the Safety Case 240
16.3.1 Test Strategy and Coverage 240
16.3.2 Artefact Inspection and Review 241
16.3.3 Review of Design Decisions 242
16.3.4 Management of Defects 243
16.3.5 Documenting Test Activities in the Safety Case 245
16.4 Summary 246
References 247

17 Gathering Evidence for the Safety Case 249
17.1 Evidencing Design Features 249
17.2 User Interface Evaluation 250
17.2.1 User Interface Evaluation Methods 251
17.2.2 Undertaking Usability Testing 254

17.3 Validating the Training Programme . . . 256
17.3.1 Validating Specific Training Controls . . . 257
17.3.2 The Training Approach as Evidence . . . 258
17.4 Validating Operating Policies . . . 259
17.5 Building Confidence in the Safety Case . . . 261
17.5.1 Historical Operation . . . 261
17.5.2 Conformance with Standards . . . 262
17.5.3 The Clinical Risk Management Workshop . . . 262
17.6 Summary . . . 263
References . . . 263

18 Developing the Clinical Safety Case . . . 265
18.1 Safety Case Structure . . . 265
18.1.1 Introduction and Executive Summary . . . 266
18.1.2 Safety Claims . . . 266
18.1.3 System Description and Scope . . . 266
18.1.4 Assurance Activities . . . 267
18.1.5 Clinical Risk Analysis and Evaluation . . . 268
18.1.6 Clinical Risk Control and Validation . . . 269
18.1.7 Outstanding Issues . . . 270
18.1.8 Conclusion and Recommendations . . . 270
18.2 The No-Go Safety Case . . . 271
18.3 Safety Case Staging . . . 272
18.4 Language and Writing Style . . . 272
18.4.1 Objectivity in Clinical Risk Management . . . 273
18.4.2 Objective Writing Style . . . 274
18.5 Summary . . . 275
References . . . 275

19 Handling Faults in Live Service . . . 277
19.1 Fault Reporting . . . 277
19.2 Assessing Faults in Live Service . . . 279
19.2.1 Describe the Problem . . . 279
19.2.2 Establish the Scale . . . 281
19.2.3 Outline the Potential Impact and Set Out the Controls . . . 282
19.2.4 Evaluate the Risk and Make Recommendations . . . 282
19.3 Escalating a Fault to the Manufacturer . . . 283
19.4 Challenges to the Assessment . . . 283
19.5 Managing Serious Faults and Incidents . . . 284
19.6 Implications for the Safety Case . . . 285
19.7 Summary . . . 286
References . . . 286

20 Maintaining the Safety Case . . . 287
20.1 Managing Change . . . 287
20.1.1 Sources of Change . . . 288

20.1.2 Detecting Change 289
20.1.3 Impacting Change 289
20.1.4 Implementing Change 290
20.1.5 Removing Functionality 291
20.2 Evolution of the Safety Case 292
20.2.1 Changes in Risk Acceptability 292
20.2.2 Learning from Operational Experience 292
20.3 Decommissioning HIT Systems 293
20.4 Summary 297
References 297

Index 299

Part I
Risk and Safety

Chapter 1
Introduction

1.1 What Is Patient Safety?

Primum non nocere, (first, do no harm) is arguably the starting point for any book on patient safety or risk management in healthcare. The origin of the phrase is disputed but scholars often cite its roots in Hippocrates' promise "to abstain from doing harm". Its premise is simple yet fundamental and powerful; in the delivery of healthcare it may be better to do nothing than to intervene and risk causing harm. Importantly primum non nocere forces us to acknowledge that even the best intentioned and well-trained clinician is capable of causing harm if our actions result in or allow a hazardous situation to develop. It is this central tenet which morally requires us to assess risk, either formally or informally, for each and every clinical activity we undertake.

Lord Darzi points out that quality in healthcare comprises three elements; safety (avoiding harm from the care that is intended to help), effectiveness (aligning care with science and ensuring efficiency) and patient-experience (including patient-centeredness, timeliness and equity) [1]. Of these, safety in the delivery of healthcare is perhaps the most fundamental expectation of our patients. In Don Berwick's report Improving The Safety of Patients in England he highlights that "patients, families and the public expect that the people and organisations that exist to help them will not hurt them. 'First do no harm' is not just a slogan in health care; it is a central aim" [2].

It is perhaps a little surprising that it is not until recent times that the term "patient safety" has become an integral part of a clinician's vocabulary. Only in the last decade have we seen healthcare organisations systematically integrate proactive risk management into the daily care of patients. And yet at the most basic level care cannot be considered to be of high quality unless it is safe [3]. Emanuel et al. define patient safety as "A discipline in the healthcare sector that applies safety science methods toward the goal of achieving a trustworthy system of healthcare delivery.

A. Stavert-Dobson, *Health Information Systems: Managing Clinical Risk*,
Health Informatics, DOI 10.1007/978-3-319-26612-1_1

Patient safety is a basic attribute of healthcare systems; it minimizes the incidence and impact of, and maximizes recovery from, adverse events" [4].

At the heart of patient safety in healthcare is the avoidance of medical error, the "failure of a planned action to be completed as intended, or use of the wrong plan to achieve an aim" [5]. Patient safety involves the study of medical error not for legal or punitive purposes but rather to help us understand the factors which contribute to adverse incidents, to enable learning and prevent harm in the future.

1.2 The History of Patient Safety

The first recognition of healthcare delivery as a cause of ill health is often attributed to Ignaz Semmelweis, a nineteenth century Hungarian physician with a passion for problem solving. Semmelweis was alarmed by the high rates of death from infection after childbirth. Through scrupulous analysis, and with no knowledge of the causative agent, he was able to identify that lack of hand hygiene amongst medical professionals and students was a significant contributor of risk. By challenging and modifying the behaviour of his colleagues he successfully demonstrated a significant reduction in mortality once hand hygiene had been addressed. Unfortunately his observations were not taken seriously by the wider medical establishment and he sadly died in an asylum in 1865 from wounds inflicted by his guards. It was only years after his death that the works of Pasteur and Lister vindicated Semmelweis's ideas.

Wind the clock forward a hundred years or so and in 1957 Chemie Grünenthal launches a new product widely thought to be the wonder drug for managing morning sickness in pregnancy. Thalidomide's antiemetic properties had mostly been discovered by chance but the drug was largely untested for its potential effects on the growing foetus. In 1961 Australian doctor William McBride wrote to the Lancet observing an increase in congenital malformations in those babies whose mothers had taken the drug. Thalidomide was subsequently withdrawn but this error of judgement in licencing the medication resulted in at least 10,000 cases of child deformity. The UK set up the Committee on Safety of Drugs in response and its modern day incarnation manages the regulation of medications to this day.

Up to this point, data on adverse incidents was largely limited to anecdote but in 1964 Schimmel [6] published an influential American paper which began to change the way healthcare providers thought. He undertook a prospective investigation of over a thousand patients and in 8 months identified 240 episodes of untoward consequences of medical care. Sixteen of these events ended in death. Later in the 1980s the Harvard Medical Practice Study [7] found that adverse event occurred in 3.7 % of hospitalisations, of those 28 % were due to negligence. 13.7 % were implicated in the patient's death. Momentum was beginning to gather and in 1984 the American Society of Anaesthesiologists established the Anesthesia Patient Safety Foundation in response to the high profile discovery of large number of anaesthetic accidents occurring during routine care.

In 1990 our understanding of the underlying causes of risk were challenged with the publication of James Reason's Human Error [8]. For the first time the role of human factors in the incident causality chain were truly characterised. He eloquently made the case for adverse events being a function not of personal inadequacies but of the environment in which individuals operate. This paved the way for transforming a largely reactive approach to risk management in healthcare to one of hazard identification and proactive risk control.

But it was in 1999 that a paradigm shift was observed. The publication of the Institute of Medicine's (IOM) To Err Is Human [5] changed the way we approach the risk of healthcare delivery forever. Extrapolating from a number of studies it concluded that up to 98,000 deaths per year in the US were due to medical error. Even using lower estimates, deaths due to medical errors exceeded the number attributable to the eighth-leading cause of death. The report highlighted that unsafe conditions resulted not from any ill-will or recklessness on behalf of individuals but from disorganised systems, fragmented care provision and a lack of identifiable drivers to improve safety. To Err Is Human succinctly concluded that the status quo was not acceptable and could not be tolerated any longer.

After To Err Is Human there was a significant rise in patient safety research [9] and general attention. A number of countries outside the US invested in research and found similar results to those highlighted in the IOM report. The percentage of hospital inpatient episodes leading to an adverse event was found to be 16.7 % in Australia [10] and 10 % in the UK [11].

The 1990s had seen a number of high profile failures in healthcare especially the widely publicised issues in Bristol's Heart Surgery service [12]. These prompted the publication of An Organisation With a Memory [13] by the UK Government's Chief Medical Officer in 2000. The report was critical of healthcare organisations' ability to learn from their mistakes and successfully implement measures to prevent a recurrence of adverse events. In particular it called for a better means of reporting incidents, a more open safety culture, a mechanism for implementing necessary changes and a wider appreciation of the value of a systematic approach.

After the millennium a number of countries successfully implemented centralised reporting tools to facilitate the capture and analysis of safety related incidents. In response to To Err Is Human the US congress enacted the Patient Safety and Quality Improvement Act of 2005 [14] to promote shared learning to enhance quality and safety nationally. This and the Patient Safety Rule [15] authorised the creation of Patient Safety Organisations to systematically collect adverse events. Analysis of this valuable data led to the detection of patterns and risks in healthcare driving strategies for mitigation.

Organisations such as the UK's National Patient Safety Agency worked tirelessly to raise awareness of patient safety, known risk factors and frameworks for the delivery of high quality care. Consistent themes included the importance of leadership at senior management level, the need for healthcare staff to be supported rather than blamed, the significance of continued learning, the sharing of good practice and the central role of the patient [2, 16]. Increasingly the term 'clinical risk' was being used to denote "the chance of an adverse outcome resulting from clinical investigation, treatment or patient care" [17].

The lack of adverse incident reporting at the start of this journey has meant that evidence of improvements in patient safety is hard to come by but progress to date has largely been slow. There have been some developments in a number of specific areas where focussed risk reduction have been implemented [18, 19]. However a report in 2011 suggested that, voluntary reporting data aside, up to one third of hospital patients still experienced an adverse medical event [20].

Organisations still frequently rely on retrospective techniques such as incident reporting and complaints to drive safety management [21] and healthcare has been slow and sporadic in adopting predictive safety assessment techniques to detect medical error [22]. Twenty two percent of people in the UK believe, when asked, that they have been the victim of medical error [23]. Nevertheless continually striving to protect patients at vulnerable points in their lives is still clearly the right thing to do and a move from reactive to proactive risk reduction will surely pay dividends in the future.

1.3 Health IT

HIT systems comprise the hardware and software that are used to electronically create, maintain, analyze, store, or receive information to help in the diagnosis, cure, mitigation, treatment, or prevention of disease [24]. Their arrival in the late 1980s began to change the way healthcare professionals deliver care or at least the manner by which information supporting clinical decisions was managed. Initially HIT systems focused on administrative activities; demographic recording, scheduling, billing and clinical orders. As the technology evolved and hardware became more affordable, the applications found roles in clinical record keeping, decision support, medication management and alerting. Aggregation of these rich data sources also facilitated secondary uses in statistical reporting, capacity management, finance and research. In some countries advances in imaging saw an almost total replacement of photographic film in radiography with the introduction of Picture Archiving and Communication Systems (PACS).

More recent trends in HIT have seen a move towards mobile technology empowering clinicians with the ability to capture, communicate and retrieve clinical data at the point of care. No longer are users tied to bulky and invariably scarce communal workstations. Advances in connectivity are also allowing patients to monitor their health at home with the intention of preventing hospital admissions and facilitating early discharge. Patients are engaging in their own care in ways that were previously impractical or impossible. By making clinical records more accessible to the patients who ultimately own them the industry is able to build a culture of healthcare transparency, trust and participation.

As well as improvements in patient safety frequently cited benefits of HIT include:

- Improved clinical outcomes
- Cost savings
- Efficiency gains
- Improved patient experience
- Enhanced decision support
- Accurate reporting of data
- Improved security of information
- Facilitation of communication
- Improved patient access to information
- Better engagement of patients in their own care

These potential advantages of HIT are significant and indeed it is the drive to realise these benefits which have given rise to the explosion in the number of HIT systems deployed. IT now represents the biggest share of many healthcare organisations' capital budgets [25]. RAND Health believe that the U.S. could make savings of more than $81 billion a year, reduce adverse incidents and improve the quality of care if HIT was widely adopted [26]. The IoM's publication Crossing The Quality Chasm [27] acknowledges that "[technology]…holds enormous potential for transforming the health care delivery". In particular, one of its key recommendations is a "renewed national commitment to building an information infrastructure" and that "This commitment should lead to the elimination of most handwritten clinical data by the end of the decade."

In reality, case studies highlight a wide range of HIT effectiveness [28]. And where benefits are shown it can be difficult to establish whether or not these have been at the expense of other healthcare objectives. Systems which improve the performance of healthcare practitioners do not automatically translate into patient outcomes [29]. Such contradictory evidence complicates the business case for HIT. A sound HIT system is just a contributor to a complex sociotechnical environment. Whether or not individual organisations successfully realise benefits depend on how the technology is integrated into care processes, the extent to which the systems are supported and the degree to which users buy-in to new ways of working. Without diligent project management and careful business transformation, the implementation of a HIT system can waste valuable resources, lower morale and ultimately put patients at risk.

1.4 HIT Impact on Patient Safety

Of all the benefits that HIT may bring, an improvement in patient safety is often high on the agenda of most commissioners. From a patient perspective any measure which reduces the likelihood of inadvertent harm occurring is invaluable. Where those systems also demonstrate efficiency gains and cost savings the business case easily develops into a thoroughly convincing argument. With such potential for human error in the delivery of care most healthcare organisations are adding HIT to their armoury. The structure and workflow that HIT systems bring has the potential to introduce consistency and predictability in care processes. Add to this the ability to revolutionise how we communicate and the development of real-time decision support, the future looks bright for this exciting technology. The Institute of Medicine advises that moving from a paper to an electronic based patient record system would be the single step that would most improve patient safety [5].

HIT can enhance patient safety in three ways: it can help prevent medical errors and adverse events; it can initiate rapid responses to any event; and it can enable the tracking of events, if they occur, and provide feedback on them to learn from [30]. But it is the first of these which is of most interest when it comes to proactively reducing the risk of harm. If we can identify the conditions in which dangerous scenarios occur we have a vital window of opportunity to intervene in the natural history of a hazard (see Sect. 2.6).

The evidence that HIT systems mitigate existing risk in the clinical environment is mixed [31, 32] and the impact on patient safety is inconclusive [33]. Most research has focussed on medication management functionality within U.S. Computerised Physician Order Entry (CPOE) systems. From the outset these applications have significant potential to influence risk in the clinical domain. Medication prescribing and administration are common healthcare activities with plenty of opportunities for harm to occur (it is estimated that there are 1.5 million preventable adverse drug events each year in the U.S. [34]). Many HIT systems provide decision support at key points in the prescribing process to augment the clinician's knowledge and professional judgement. By reducing time pressures and lessening the need for clinicians to hold an encyclopaedic knowledge of the medicines they prescribe, one would expect a reduction in the incidence of slips, lapses and mistakes (see Sect. 5.3).

In one study by Eslami researchers systematically analysed 30 HIT related articles to assess the impact of CPOE systems in hospital out-patient departments [35]. Only one study demonstrated a significant reduction in adverse medication events. Eslami summarised that the relatively small number of evaluation studies published to date do not provide adequate evidence that CPOE systems enhance safety and reduce cost in the outpatient settings. In contrast a literature review by Ammenwerth found that 23 out of 25 studies showed a significant risk reduction in hospital medication errors [36]. He concludes that electronic prescribing seems to be a useful intervention for reducing the risk of medication errors and adverse events. Similarly in the UK researchers at the Wirral Hospital NHS Trust found that the introduction of HIT-supported medication pathways significantly reduced errors in the prescription of some high-risk drugs [37]. Encinosa reported a reduction of adverse drug incidents by 69 % in some hospitals [38].

HIT is finding an increasing role in healthcare delivery and no longer is its application limited to simple order entry. Outside of medication management the evidence for HIT playing an active role in the reduction of risk is lacking and much more research is required [39]. This is hampered by no obvious or consistent benchmark by which to measure these benefits. Nevertheless, the prospect of HIT better informing care delivery makes it likely that in the near future improvements in patient safety will be achieved. The US Office of the National Coordinator for

Health IT believes that "the potential for health IT systems to dramatically improve patient safety is within our reach" [40].

What is perhaps more difficult and even awkward to acknowledge is the notion that HIT plays its own part in the introduction of risk to the clinical environment. In the early days of HIT, functionality was limited to simple administrative tasks or the straight-forward archiving and retrieval of basic clinical data. In these systems the underlying logic was transparent and discernible with any decision making remaining entirely in the hands of clinicians. Once systems began to provide time efficiencies, improved methods of communication and in-the-moment decision support, healthcare staff naturally began to rely on such valuable tools. But when those tools let us down hazardous situations can quickly develop. Like drugs 50 years ago, products in health informatics are largely unregulated with regard to safety [41].

Systems which are poorly designed or sub-optimally implemented can add unnecessary complexity to the clinical environment. It is these ideas that lead Weiner and his colleagues [42] to coin the term e-iatrogenesis. They define this simply as "patient harm caused at least in part by the application of health information technology". In this way two key challenges are presented by HIT; one of security (who has access to which information?) and one of safety (can we trust the information we use for critical decisions?) [43]. It is the latter which is of primary focus in this book.

The mitigation of risk through the implementation of HIT needs to be carefully balanced with that which is introduced. Interestingly this exchange is not necessarily like for like. A pre-existing hazard in the clinical environment is likely to be well characterised and understood. For example, clinicians are aware that transcribing medications onto or between paper drug charts is an activity known to be associated with risk. Even the busiest clinician typically carries out the process with the care and attention that it deserves.

A HIT system designed to assist in this process may bring with it some idiosyncrasies. These behaviours are likely to be less familiar, poorly characterised and less well understood by its users. Without effective management the overall risk picture could deteriorate, at least in the short term. The Committee on Quality of Health Care in America concludes that health care organisations "should expect any new technology to introduce new sources of error and should adopt the custom of automating cautiously, alert to the possibility of unintended harm" [5].

In Ash's study into the unintended consequences of CPOEs [44] he and his team identified nine categories of issues introduced by HIT (Table 1.1). If we accept that the number of HIT systems being deployed is rising rapidly then assuming a fixed probability of harm associated with each system, there can indeed only be increased harm if the situation remains unchecked [45]. Systems tend to exhibit inertia and change slowly – the ability to implement effective controls (i.e. measures to strategically reduce the risk) can often lag behind the drive to deploy new innovative systems.

The academic literature and popular media contain numerous other examples of HIT contributing to patient harm. Probably the most serious accident of this type was attributable to the Therac-25 radiation therapy machine in the 1980s. Six patients in the US were subject to massive doses of radiation resulting in serious

Table 1.1 Nine categories of unintended consequences of CPOEs in decreasing order of frequency

1. Rework required by clinicians
2. Workflow problems
3. Constant demands for technology refresh
4. Continued use of paper
5. Failure to communicate information
6. Negative emotions of users
7. New kinds of errors
8. Enforced clinical practices
9. Over-dependence on the technology

injury and death. The problem was down to a series of undetected design issues in the product's software components. An investigation conducted by Leveson in 1993 [46] led to changes in the way manufacturers and regulators handle software design in medical devices. The findings highlighted a number of shortcomings which interestingly shone a light on some of the cultural practices at the time:

- Placing excessive confidence in the integrity of software to the extent that the manufacturer removed some hardware safety features which had been present in previous versions of the product.
- Failure to thoroughly examine and safety assess the software compared to the hardware.
- An early assumption that the problem was due to a particular hardware fault (therefore largely ignoring the underlying software issues) despite there being little evidence to support this.
- A lack of features to independently monitor the dose of radiation being delivered and instead trusting information reported by the software (which in this case was incorrect).
- Less than acceptable safety engineering procedures (for example, a lack of documentation and over-reliance on a single programmer being responsible for building the software).
- Overly optimistic safety assessments and over-confidence that fixes had successfully rectified the underlying fault
- Lack of incident reporting mechanisms and failure of the manufacturer to fully investigate the adverse events as they arose.

The Therac-25 incident is perhaps an extreme example; most HIT is not responsible for the automated delivery of drugs, radiation or other forms of energy. Nevertheless these events highlight the potential for sub-optimally engineered systems or the way they are operated to cause significant harm to those whom we have a duty of care.

Other examples of notable incidents in the press include:

- An incident whereby a patient was issued with a barcoded wrist band which was erroneously associated with a different individual's electronic record [47]. The patient was incorrectly treated for hyperglycaemia and administered a potentially fatal dose of insulin.

- In a 2011 US case, a pharmacy technician entered incorrect data into an electronic form which resulted in a baby being administered a lethal infusion which contained 60 times the intended amount of sodium chloride [48].
- In 2014 a case was reported in the UK of a child with congenital heart disease who failed to receive an important follow-up appointment and echocardiogram after undergoing heart surgery [49]. The error was due to an issue with the hospital's appointment booking system and the child died in September 2012. The coroner indicated that "Due to the failure of the hospital's outpatient booking system, there was a 5 month delay in [the child] being seen and receiving necessary treatment."
- An incident in 2007 in the U.S. Dominican Hospital resulted in a patient undergoing an unnecessary appendectomy [50]. Two female patients had undergone diagnostic CT scans but an incompatibility between the computer systems in the scan room and that used by the radiologist contributed to the scans being associated with the wrong patients. The state Department of Public Health fined Dominican Hospital $25,000.

Case studies aside, the true incidence of adverse events occurring as a result of HIT implementation is ill-defined. This is perhaps due to inconsistencies in reporting and the practical difficulties of defining the precise contributory role of the system. Again most studies have focussed on CPOE systems in the U.S. For example Koppel et al. [51] surveyed 261 users of a CPOE system in a large teaching hospital. It was found that the application facilitated 22 different types of medication error. These included fragmented views of medication screens and inventory displays being mistaken for dosage guidelines. In 2008 the US Joint Commission issued an alert regarding technology-related adverse events. They reported that 25 % of the medication incidents logged on the United States Pharmacopeia MEDMARX database in 2006 involved some aspect of computer technology [52].

In one study undertaken at a children's hospital, Han et al. [53] observed a doubling in patient mortality which coincided with an implementation of a CPOE system introduced in an attempt to reduce clinical risk. Interestingly two other studies were carried out on different implementations of the same system. One showed no impact on mortality [54] whilst the other showed some reduction [55] perhaps suggesting that factors relating to the deployment and implementation of the system rather than the application itself were at play.

A number of studies have made attempts to characterise the kinds of incidents which occur along with their relative frequency and impact on patients. In 2012 the US ECRI Institute asked participating healthcare organisations to submit at least ten HIT related incidents using a patient safety event reporting system [56]. This resulted in 171 incidents across 36 facilities. Sixty one percent of events failed to be detected prior to reaching the patient, 23 % were classified as near-misses. Six percent of the 171 incidents contributed to or resulted in actual harm. More recently in 2015 Magrabi et al. examined 850 safety events which had been reported during England's National Programme for IT (see Sect. 4.2.2) between 2005 and 2011

Table 1.2 Adverse incidents by problem type (Magrabi et al. [58])

Clinical problem type	%
Medication problems	41
Clinical process problems	33
Exposure to radiation	15
Surgery problems	11

[57]. Of these 24 % had an observable impact on care, 4 % were categorised as near-misses and 3 % resulted in actual patient harm (including three deaths).

In terms of the types of HIT systems involved, CPOE systems were implicated the most in the study (25 % of events) followed by clinical documentation systems (17 %) and electronic medication administration records (15 %). Fifty three percent of the incidents were associated with IT systems involved in medication management. Each event was categorised according to the causative agent using an established classification system. Fifty five percent of the events were directly related to the system itself whilst 45 % were as a result of human interaction with the application. Of those related directly to the system, issues with the system interface (28 %) were the most common followed by system configuration problems (23 %). Of those classified as relating to human interaction users entering the wrong data was the most common incident (32 %) followed by the wrong patient record being retrieved (25 %). Only 3 % of incidents occurred as a result of system unavailability.

In 2011, Magrabi analysed 46 adverse events which had been submitted to the US Food and Drug Agency (FDA) [58] and classified them by problem type (Table 1.2). Ninety three percent involved CPOE or PACS systems.

Medication problems included misidentification, wrong medication, wrong dose and missed doses. In one case a CPOE system "required users to scroll through 225 options on a drop down menu. The options were arranged in a counterintuitive alphabetical order, and resulted in a patient being overdosed with four times more digoxin than required." One in three events were associated with clinical process problems. For example, entry of a portable x-ray image into a PACS system under the wrong name resulted in "a wrong diagnosis and subsequent intubation which may have contributed to death".

In summary the research shows that CPOE and medication management systems in particular, by no means exclusively, are capable of introducing clinical risk. The underlying causes are multifactorial and whilst system unavailability is not a significant contributor both the technology and the way it is operated can result in harm.

1.5 Controlling Risks from HIT

Managing the safety of drugs depends on controlled trials which are not realistically feasible or affordable as a mandatory requirement for HIT [41]. The safety of HIT systems is currently controlled through a mixture of legislation and administration depending on the nature of the product and the country of implementation and manufacture. The result is a rather complex web of regulation, procurement rules and compliance with local and international standards. Products at the higher end of the risk spectrum are often designated as medical devices and as such manufacturers need to comply with formal regulatory requirements. However, it is not uncommon for these frameworks to lag behind the technology and at times the legislative boundaries are somewhat blurred. Many HIT systems considered to be non-medical devices are still capable of adversely impacting care and some authorities have made a case for all HIT being subject to formal regulatory control [41]. Currently, outside of medical device regulation, formal requirements for risk management are not well established. Manufacturers and their customers are left without a clear way forward yet are potentially legally exposed.

Suppliers of HIT systems have been slow to adopt approaches to safety management outside of medical device regulation. In an increasingly litigious society a vendor's acknowledgement that their application can introduce a risk to patients can be erroneously perceived as admitting liability. In many cases suppliers attempt to manage risk through the use of disclaimers and waivers in an attempt to transfer liability onto users or their employers. When it comes to disclosing HIT related adverse incidents suppliers often contractually limit their customer's ability to publically report and share issues, constraints which are somewhat counterproductive to a safety culture [59]. The IoM in their report on Building Safer Systems for Better Care acknowledge that "Currently, there is no systematic regulation or sense of shared accountability for product functioning, liability is shifted primarily onto users, and there is no way to publicly track adverse outcomes." [60] Commentators have made calls for self-assessment, certification and even periodic inspections of electronic health record manufacturers [61].

In the drive to protect intellectual property suppliers are often reluctant to divulge details of their product's inner workings to healthcare organisations. This leaves them with the difficult challenge of convincing their customers that their product is safe without explaining the rationale that has lead them to that conclusion. In the absence of any official reporting mechanisms, healthcare organisations sometimes seek solace in providing informal risk management support to each other through user groups, social media and other forums. Whilst this approach may be a first step towards mutual learning it can hardly be relied upon as a systematic strategy. Occasionally even informal discussions can be constrained by non-disclosure agreements and other contractual tools which limit the active sharing of information.

This situation is all in stark contrast to other safety critical industries. For example in aerospace there is a more open and transparent culture of learning which crosses organisational and otherwise commercially competitive boundaries. In aerospace stakeholders are aware that improvements in safety are for the greater good of the industry and that in the long term a safe product drives revenues for all players. HIT suppliers have some way to catch up and it is largely the responsibility of their customers to call for the transparency that is required to enable a rigorous and practical assurance process.

1.6 Regulation of Health IT Systems

Regulation imposes specific requirements on HIT manufacturers to meet defined quality criteria and address clinical risk – this is no easy task and can require significant investment. The regulations are typically woven into law and in many cases, failure to comply is a criminal offence.

HIT is a mixed bag of technology with wide variation in functionality, ability to influence clinical decision making and risk to patients. The effort expended in assuring a system needs to be commensurate with the risk it poses. Manufacturers have finite resources and would not wish to subject all technology to the same level of analysis and rigour. The cost and effort involved would simply not be justified in many cases and regulators are generally sympathetic to this position. Instead these organisations attempt to classify products according to risk reserving the greatest level of analysis for those products which have the most potential to adversely impact care should they malfunction.

HIT systems can broadly be categorised into four classes:

- Business management systems
- Care administration systems
- Clinical management systems
- Medical devices

Whilst perhaps and over simplification, in general the ability of the system to influence care and therefore the potential risk to patients increases down the list. Nevertheless one should not assume that just because a system seems to fit a specific class then a particular approach to risk management is warranted or can be or justified. The level of analytical rigour and need for risk reduction depends entirely on a particular system's potential to adversely impact care irrespective of how one might choose to classify it.

1.6.1 Business Management Systems

These systems are associated with the running and operation of the healthcare service. Often the applications are not specific to health and sometimes they are general tools which have been configured for use in a healthcare setting. Typical systems are those which support business functions such as:

- Payroll and Human Resources
- Staff rostering
- Finance management
- Billing
- Complaints
- Asset and inventory management
- Statistical reporting

- Simple telephony
- General tools such as word processors, spreadsheets, etc.

In the main systems of this nature do not influence care sufficiently to introduce any significant clinical risk (or at least to influence care is not part of their intended purpose). Where dependencies are likely to exist it would appropriate to consider clinical risk – otherwise the focus of any analysis will be on the business. As such most countries choose not to regulate these kinds of application in healthcare.

Note though that it is nearly always possible through inference to derive scenarios where failures of these systems could eventually compromise patient care. For example, errors in staff rostering could contribute to a service being understaffed which could then lead to an adverse outcome. Generally the harm causality pathway in these cases is tenuous or convoluted. There is sufficient human intellect embedded within the business processes to catch potential problems long before they realistically affect care. In other words any hazards are simply not credible in the real world or would fall into a category where risk reduction would not be justified by the effort required.

For organisations looking to manufacture or implement these systems though it is wise to preserve the rationale for choosing not to undertake clinical risk management activities. This demonstrates to any regulator, auditor or indeed customer that due consideration of the potential for the technology to impact care has been made. Future generations of personnel are also able to understand the reasons and context behind historical decisions should circumstances change.

1.6.2 Care Administration Systems

Care administration systems deal with the practical organisation of care delivery but stop short of capturing and delivering the information on which clinical decisions are made.

Systems in this class might include:

- Patient identification services
- Admission, discharge and transfer management
- Scheduling services for appointments, surgical procedures, etc.
- Referral management
- Bed management
- Pharmacy stock control
- Clinic management

Care administration systems have the capability to more directly influence the provision of and access to care. Their lack of ability to affect clinical decision making in no way precludes them from being safety-related. In fact significant risk can exist in relying on these system to manage care delivery. For example, suppose a system fails to create appointment letters at the appropriate time such that patients fail to turn up at a clinic. The purely administrative nature of this hazard on no account diminishes its potential to cause harm.

In most countries these systems tend not to be subject to formal regulation. Nevertheless their ability to impact care means that their manufacture and implementation will benefit from clinical risk management. This will typically take the form of voluntary self-assessment or might be a requirement in a commercial contract or procurement rule.

1.6.3 Clinical Management Systems

Clinical management systems form the heart of care delivery and decision making. These applications capture, retrieve and communicate clinical data to inform clinicians about the ongoing care of individual patients. At the simplest level they act like the paper notes historically used for documenting clinical activities although often they do much more. Clinical management systems include:

- Clinical documentation and noting
- Order communication and diagnostic results management
- Electronic prescribing and drug administration
- Simple decision support
- Radiology information systems
- Immunisation management
- Specialist clinical data capture, e.g. maternity record, etc.

Failure of these systems has the potential to immediately impact clinical care. At the time of writing most clinical management systems are not medical devices and as such manufacturers currently enjoy the relative freedom to innovate. These systems are complex however and have to be dependable and as a result they benefit significantly from being the subject of formal safety assurance. Increasingly manufacturers and healthcare organisations are being encouraged by governments and industry bodies to include risk management in their lifecycles. Organisations can usually decide for themselves how best to assess and manage clinical risk. Providing the approach is transparent, logical and undertaken with a level of rigour commensurate with the risk then generally organisations are free to justify safety claims in a way that works for their stakeholders.

1.6.4 Medical Devices

The precise definition of a medical device varies depending on the regulatory framework under consideration and even interpretation by individual countries. Examples include the European Commission's Medical Device Directive (MDD) and that of the US Food and Drug Administration. Whilst in principle each country can regulate differently, some governments choose to base their regulations and standards on those of other countries. Regulators have attempted to focus on those products at the higher end of the clinical risk spectrum, i.e. those with the greatest potential for harm should they function incorrectly.

Medical devices are usually classified further depending on risk [62], for example the MDD has four classes with increasingly rigorous assurance activities needed to comply. Medical device legislation has its origins in controlling physical products such as implantable and invasive devices. Applying the same principles to computer software can require a level of translation and some regulators have been helpful in providing guidance on this [63].

Interestingly whether a product qualifies as a medical device is generally not dependent on the outcome of the regulatory risk assessment. Instead classification is usually based on the product's intended purpose (essentially taken to be a high level analogue of risk). Thus, classification occurs first depending on what the product is to be used for and then formal risk management is undertaken as part of the compliance process. Similarly in the US the focus of FDA compliance often comes down demonstrating equivalence to a similar product rather than declaring the outcome of an objective risk assessment.

There is variability between regulators on the characteristics of software systems which constitute being a medical device and the Global Harmonization Task Force has gone some way to standardising definitions [64]. In practice HIT systems which fall into medical device regulation typically include:

- Devices with a measuring function such as vital signs monitors and glucometers
- Applications which produce new clinical data from calculations particularly where it is not obvious to the end user how that data has been derived.
- Systems which make use of embedded algorithms and provide advice based on information and decisions which are not wholly transparent to the user
- Applications with a real time monitoring or alarm function used to assess the second by second health of a patient
- Systems which provide the primary or sole means by which a clinical diagnosis is made
- Software which is used to drive the delivery of ionising radiation or other invasive interventions

Software applications which are generally classified as medical devices include:

- Picture Archiving and Retrieval Systems (PACS)
- Clinical calculators
- Chemotherapy systems
- Advanced decision support applications
- Systems tightly integrated with point of care testing devices

What systems in this class have in common is that they are performing functions beyond simple archiving and retrieval of information. The information provided by these systems is often the sole determinant of a patient's immediate care. Essentially these systems own a considerable share of the decision making power and should they malfunction there can be a significant risk of patient harm.

Medical devices need to be assured in a manner which is acceptable to a regulator and the requirements are usually embodied in law. Regulatory conformance often involves being compliant with one or more standards linked to the local regulatory framework. Paradoxically providing assurance can in some ways be easier for medical devices as the risk and quality management expectations are clearly set out and well understood by practitioners.

1.6.5 Health Applications Aimed at Mobile Devices

In recent years the trend for software to be moved to mobile platforms has opened up opportunities for commodity applications which monitor our well-being and provide health advice. Many hundreds of these now exist and range from carefully designed medical devices capable of monitoring complex medical conditions to poorly crafted programs offering little more than quackery.

Health apps can largely be categorised as follows:

1. Mobile health information systems aimed at clinicians
2. Consumer facing disease management products
3. Consumer facing wellness and lifestyle products

Apps which are designed for healthcare professionals are essentially extensions of traditional Electronic Health Records (EHRs) but often with narrowed or very specific functionality. These can be considered clinical management systems or medical devices depending on the intended purpose. From a patient safety perspective, risk management is required in exactly the same way as for conventional health information systems if they are able in some way to adversely impact care.

When instead apps are aimed at the consumer market, risk management responsibilities move much further towards the manufacturer. Without any healthcare professional or healthcare delivery organisation in the information chain the manufacturer shoulders full responsibility for design, configuration and maintenance. The UK British Standards Institute in 2015 issued a publication known as PAS 277 to provide advice on how quality management might work in these circumstances [65]. It specifically calls for manufacturers to include risk management in the product lifecycle but stops short of specifying a particular methodology.

When it comes to consumer facing lifestyle technology the question is as to whether software which promotes wellness rather than diagnosing or treating a medical condition could itself contribute to harm. For example if a product provides advice on weight management and the guidance offered turned out to be incorrect would this constitute harm? Similarly, would a fitness app be harmful if it were to recommend a level of exercise which was in fact dangerous?

The need for regulation and risk management in these circumstances is subject internationally to a great deal of debate. In 2015 the US FDA provided some draft guidance [66]. It defines a general wellness product as one which:

…has (1) an intended use that relates to a maintaining or encouraging a general state of health or a healthy activity, or (2) an intended use claim that associates the role of healthy lifestyle with helping to reduce the risk or impact of certain chronic diseases or conditions and where it is well understood and accepted that healthy lifestyle choices may play an important role in health outcomes for the disease or condition).

Products which meet the definition will not be required to meet any specific regulatory requirements in the US. Nevertheless, where a product fails to meet the inclusion criteria for being a medical device manufacturers may be wise to consider a formal risk assessment. Doing so would demonstrate that due consideration had been given to any potential harm which might result from its use. This could have implications for the way the product is described and the content of the instructions for use. Even if a risk assessment concluded that no harm could occur from the product's operation, the diligence applied is likely to sit well in court should there be any future legal claim against the manufacture.

1.7 Summary

- Too many patients are unintentionally harmed in the course of receiving healthcare and progress has been slow in finding solutions.
- HIT has an important role to play in controlling clinical risk but when systems are poorly designed or sub-optimally implemented, risk can be introduced as well as mitigated.
- Manufacturers and healthcare organisations have a responsibility to proactively manage clinical risk rather than react only when adverse events occur.
- Not all forms of HIT introduce risk to the same extent and the means by which products are assured needs to take account of the degree to which they could cause harm.
- For non-medical devices the answer to better clinical risk management is not additional regulation but rather an improved understanding of how organisations can efficiently and effectively execute self-assessment and practical risk mitigation.

References

1. UK Department of Health. High quality care for all. NHS next stage review final report. 2008.
2. National Advisory Group on the Safety of Patients in England. A promise to learn – a commitment to act. Improving the safety of patients in England. Br Med J. 2013. https://www.gov.uk/government/uploads/system/uploads/attachment_data/file/226703/Berwick_Report.pdf.
3. Moss F, Barach P. Quality and safety in health care: a time of transition. Qual Saf Health Care. 2002;11:1.
4. Emanuel L, Berwick D, Conway J, Combes J, Hatlie M, Leape L, et al. Advances in patient safety: new directions and alternative approaches (vol. 1: assessment). 2008. http://www.ahrq.gov/professionals/quality-patient-safety/patient-safety-resources/resources/advances-in-patient-safety-2/index.html#v1.

5. Kohn L, Corrigan J, Donaldson M. To err is human: building a safer health system. Washington, DC: National Academies Press. Committee on Quality of Health Care in America, Institute of Medicine; 2000.
6. Schimmel E. The hazards of hospitalization. Ann Intern Med. 1964;60:100–10.
7. Brennan T, Leape L, Laird N, Hebert L, Localio A, Lawthers A, et al. Incidence of adverse events and negligence in hospitalized patients: results of the Harvard Medical Practice Study. Qual Saf Health Care. 2004;13:145–52.
8. Reason J. Human error. 1st ed. Cambridge: Cambridge University Press; 1990.
9. Stelfox H, Palmisani S, Scurlock C, Orav E, Bates D. The "To Err is Human" report and the patient safety literature. Qual Saf Health Care. 2006;15(3):174–8.
10. Wilson R, Runciman W, Gibberd R, Harrison B, Newby L, Hamilton J. The quality in Australian Health Care Study. Med J Aust. 1995;163:458–71.
11. Vincent C, Neale G, Woloshynowych M. Adverse events in British hospitals: preliminary retrospective record review. BMJ. 2001;322:517–9.
12. Smith R. Regulation of doctors and the Bristol inquiry. Both need to be credible to both the public and doctors. BMJ. 1998;317:1539–40.
13. UK Department Of Health. An organisation with a memory. Report of an expert group on learning from adverse events in the NHS chaired by the Chief Medical Officer. 2000.
14. Patient Safety And Quality Improvement Act. Public Law 109–41. 109th Congress; 2005.
15. Department of Health and Human Services. Patient safety and quality improvement; final rule. 2008.
16. NHS National Patient Safety Agency. Seven steps to patient safety, the full reference guide. 2004.
17. National Patient Safety Agency. Healthcare risk assessment made easy; 2007.
18. de Vries E, Prins H, Crolla R, den Outer A, van Andel G, van Helden S, et al. Effect of a comprehensive surgical safety system on patient outcomes. N Engl J Med. 2010;363:1928–37.
19. Berenholtz S, Pronovost P, Lipsett P, Hobson D, Earsing K, Farley J, et al. Eliminating catheter-related bloodstream infections in the intensive care unit. Crit Care Med. 2004;32(10):2014–20.
20. Classen D, Resar R, Griffin F, Federico F, Frankel T, Kimmel N, et al. 'Global trigger tool' shows that adverse events in hospitals may be ten times greater than previously measured. Health Aff. 2011;30(4):581–9.
21. Anderson J, Kodate N, Walters R, Dodds A. Can incident reporting improve safety? Healthcare practitioners' views of the effectiveness of incident reporting. Int J Qual Health Care. 2013;25:141–50.
22. Lyons M. Towards a framework to select techniques for error prediction: supporting novice users in the healthcare sector. Appl Ergon. 2009;40(3):379–95.
23. Schoen C, Osborn R, Trang Huynh P, Doty M, Zapert K, Peugh J, et al. Taking the pulse of health care systems: experiences of patients with health problems in six countries. Health Aff – Web Exclusive. 2005;W5-509-25.
24. Agency for Healthcare Research and Quality. Device or medical/surgical supply, including health information technology (HIT). In: Hospital common formats—version 1.2: event descriptions, sample reports, and forms. 2013.
25. Ash J, Berg M, Coiera E. Some unintended consequences of information technology in health care: the nature of patient care information system-related errors. J Am Med Inform Assoc. 2004;11(2):104–12.
26. RAND Corporation. Health information technology: can HIT lower costs and improve quality? 2005.
27. Institute of Medicine. Crossing the quality chasm: a new health system for the 21st century. 2001.
28. Chaudhry B, Wang J, Wu S, Maglione M, Mojica W, Roth E, et al. Systematic review: impact of health information technology on quality, efficiency, and costs of medical care. Ann Intern Med. 2006;144(10):742–52.
29. Garg A, Adhikari N, McDonald H, Rosas-Arellano M, Devereaux P, Beyene J, et al. Effects of computerized clinical decision support systems on practitioner performance and patient outcomes: a systematic review. JAMA. 2005;293(10):1223–38.
30. Stroetmann V, Thierry J, Stroetmann K, Dobrev A. eHealth for safety, impact of ICT on patient safety and risk management. 2007.

31. Metzger J, Welebob E, Bates D, Lipsitz S, Classen D. Mixed results in the safety performance of computerized physician order entry. Health Aff. 2010;29(4):655–63.
32. Banger A, Graber M. Recent evidence that health IT improves patient safety. 2015.
33. Hydari M, Telang R, Marella W. Saving Patient Ryan. Can advanced electronic medical records make patient care safer? Social Science Research Network; 2014.
34. Institute of Medicine. Preventing medication errors: quality chasm series. 2006.
35. Eslami S, Abu-Hanna A, deKeizer N. Evaluation of outpatient computerized physician medication order entry systems: a systematic review. J Am Med Inform Assoc. 2007;14:400–6.
36. Ammenwerth E, Schnell-Inderst P, Machan C, et al. The effect of electronic prescribing on medication errors and adverse drug events: a systematic review. J Am Med Inform Assoc. 2008;15(5):585–600.
37. Farrar K, Caldwell N, Robertson J, et al. Use of structured paediatric-prescribing screens to reduce the risk of medication errors in the care of children. Br J Healthc Comput Inf Manag. 2003;20:25–7.
38. Encinosaa W, Bae J. Meaningful use IT reduces hospital-caused adverse drug events even at challenged hospitals. Healthcare. Aug 2014.
39. Black A, Car J, Pagliari C, Anandan C, Cresswell K, Bokun T, et al. The impact of eHealth on the quality and safety of health care: a systematic overview. Public Libr Sci Med. 2011;8(1):e1000387.
40. US Office of the National Coordinator for Health IT. ONC Health IT safety program – progress on health IT patient safety action and surveillance plan. 2014.
41. Rigby M, Forsström J, Roberts R, Wyatt J. Verifying quality and safety in health informatics services. BMJ. 2001;323:552.
42. Weiner J, Kfuri T, Chan K, Fowles J. "e-Iatrogenesis": the most critical unintended consequence of CPOE and other HIT. J Am Med Inform Assoc. 2007;14:387–8.
43. Stalhane T. The role of HazOp in testing health-care applications. In: 2005; 5th conference on software validation for health care. Congress Center Stadthalle Düsseldorf, Germany: The Norwegian University.
44. Ash J, Sittig D, Dykstra R, Campbell E, Guappone K. The unintended consequences of computerized provider order entry: findings from a mixed methods exploration. Int J Med Inform. 2009;78 Suppl 1:S69–76.
45. Coiera E, Aarts J, Kulikowski C. The dangerous decade. J Am Med Inform Assoc. 2012;19(1):2–5.
46. Leveson N, Turner C. An investigation of the Therac-25 accidents. IEEE Comput. 1993; 26(7):18–41.
47. McDonald C. Computerization can create safety hazards: a bar-coding near miss. Erratum Ann Intern Med. 2006;144(7):510–6.
48. Graham J, Dizikes C. Baby's death spotlights safety risks linked to computerized systems. Chicago Tribune. 27 Jun 2011.
49. The Bristol Post. Samuel starr inquest: coroner blames "failure" of NHS computer system for boy's death. 5 Mar 2014.
50. California Department of Public Health. Statement of deciciencies and plan of correction. 2010.
51. Koppel R, Metlay J, Cohen A, Abaluck B, Localio A, Kimmel S, et al. Role of computerized physician order entry systems in facilitating medication errors. JAMA. 2005;293(10):1197–203.
52. Sentinel event alert. Safely implementing health information and converging technologies. Jt Comm. Dec 2008;11:(42). http://www.jointcommission.org/assets/1/18/SEA_42.PDF.
53. Han Y, Carcillo J, Venkataraman S, et al. Unexpected increased mortality after implementation of a commercially sold CPOE system. Pediatrics. 2005;116:1506–12.
54. Del Beccaro M, Jeffries H, Eisenberg M, Harry E. Computerized provider order entry implementation: no association with increased mortality rates in an intensive care unit. Pediatrics. 2006;118(1):290–5.
55. Longhurst C, Parast L, Sandborg C, Widen E, Sullivan J, Hahn J, et al. Decrease in hospital-wide mortality rate after implementation of a commercially sold computerized physician order entry system. Pediatrics. 2010;126(1):14–21.

56. ECRI Institute. ECRI Institute PSO Deep Dive: Health Information Technology. ECRI Institute PSO; 2012.
57. Magrabi F, Baker M, Sinha I, Ong M, Harrison S, Kidd M, et al. Clinical safety of England's national programme for IT: a retrospective analysis of all reported safety events 2005 to 2011. Int J Med Inform. 2015;84(3):198–206.
58. Magrabi F, Ong M, Runciman W, Coiera E. Patient safety problems associated with healthcare information technology: an analysis of adverse events reported to the US Food and Drug Administration. AMIA annu symp proc. 2011; p. 853–57.
59. Koppel R, Kreda D. Health care information technology vendors' "hold harmless" clause: implications for patients and clinicians. JAMA. 2009;301(12):1276–8.
60. Institute of Medicine. Health IT and patient safety: building safer systems for better care. 2012.
61. Sittig D, Classen D. Safe Electronic Health Record use requires a comprehensive monitoring and evaluation framework. JAMA. 2010;303(5):450–1.
62. European Commission. Medical device: guidance document. Classification of medical devices. Guidelines relating to the application of the Council Directive 93/42/EEC on Medical Devices. 2010.
63. European Commission. Guidelines on the qualification and classification of stand alone software used in healthcare within the regulatory framework of medical devices. 2012. Report No.: MEDDEV 2.1/6.
64. Global Harmonization Task Force. Definition of the terms 'Medical Device' and 'In Vitro Diagnostic (IVD) Medical Device'. Study Group 1 of the Global Harmonization Task Force; 2012.
65. British Standards Institute. PAS 277:2015. Health and wellness apps – quality criteria across the life cycle – code of practice. 2015.
66. Food and Drug Administration. General Wellness: policy for low risk devices. Draft guidance for industry and food and drug Administration Staff. U.S. Department of Health and Human Services; 2015.

Chapter 2
Risk and Risk Management

2.1 Overview of Risk

Risk is an inevitable part of life. Indeed some personality types would argue that the very taking of risk itself enhances life and provides fun, excitement and opportunities. Remove the risk and the element of danger somehow lessens the experience. A decision to drive to the shops, eat a fatty meal, smoke a cigarette or even cross the road involves an acceptance of risk based on the perceived benefit the activity offers against the threat it poses. We may like to think that those decisions are rational and evidence based.

In reality our choices are frequently built on our perception of risk which is subject to extensive cognitive bias. We are influenced by the behaviour of those around us, the popular media, our past experiences and superstitions. Lowrance [1] notes the following "Nothing can be absolutely free of risk. One can't think of anything that isn't, under some circumstances, able to cause harm. Because nothing can be absolutely free of risk, nothing can be said to be absolutely safe. There are degrees of risk and, consequently, there are degrees of safety."

When it comes to healthcare delivery, as Don Berwick acknowledges [2]. "Even though hazards in care cannot be eliminated, harm to patients can be and should be reduced continually, everywhere and forever. The fight for safety is a never-ending struggle against entropy, engaged tirelessly and with focus against an enemy that continually emerges and re-emerges" [2]. Note that in this book the term harm is used generically. In some hazardous scenarios harm may fail to arise purely by chance, a so called 'no harm event' [3]. For simplicity these situations are not distinguished.

The Oxford English Dictionary defines risk as "[Exposure to] the possibility of loss, injury, or other adverse or unwelcome circumstance; a chance or situation involving such a possibility" [4]. The focus here, as with a number of other definitions is that risk involves the loss of something we value. In the context of Clinical Risk Management (CRM) this is likely to be our health, wellbeing or indeed our

A. Stavert-Dobson, *Health Information Systems: Managing Clinical Risk*,
Health Informatics, DOI 10.1007/978-3-319-26612-1_2

life. The definition also helpfully introduces the idea that loss is not something which is inevitable or guaranteed but rather something which may or may not happen depending on the precise circumstances involved. This potentially provides us with the opportunity to intervene in the events leading to loss and thereby control the risk.

Another perspective on risk is offered by ISO 31000 [5] where it is defined as the "effect of uncertainty on objectives". In other words, if we accept that achieving our aims requires the successful completion of a series of discrete steps then risk represents the possibility that, at any of those steps, a lack of knowledge may cause us to deviate in a way that prevents us achieving our goal. The introduction of uncertainty as a fundamental basis for risk has some important implications particularly in CRM. One might argue that, on the back of this definition, if we had sufficient knowledge about a system then risk could be eliminated or at least be eminently manageable through the application of simple logic. However we have to consider uncertainty in a broader sense particularly in the ability to predict the actions of a system's users, their operating environment, motivation and objectives. As we will later go on to discuss, uncertainty and unpredictability in these areas can represent significant sources of risk.

Risk in an inherent part of industry, manufacturing and engineering whether it be risk to an organisation's employees, the users of the products manufactured or the general public. A number of practice areas have grown up with the concept of risk and safety at the heart of operations. These so called High Reliability Organisations (HROs) include aviation, nuclear engineering, oil drilling, chemical engineering and the military. These industries carry significant risk in their day to day operation. Society needs to carefully balance the benefits of these products and services against the risk they pose whilst the industry has to balance risk against investment. Effective risk management sits at the core of these industries in order to avert the constant threat of potential disaster and loss of life.

Whilst the risk posed by HIT is generally not of the same magnitude, the growing proliferation of technology in healthcare delivery means that even minor shortcoming in clinical systems have the potential to impact large numbers of patients over a period of time. Many of the approaches to effective CRM have been inherited from those operated by HROs. Whilst the techniques are not always a perfect fit, the HIT industry has much to learn from the seemingly disconnected worlds of aviation and military hardware.

2.2 Types of Risk

So far we have considered risk as a component of safety. However, there are many other uses of the term. For example, one will find references to business risk, environmental risk, financial risk and reputational risk. It is important to realise that these types of risk are different and distinct from clinical risk. For example, in the financial sector risk is expressed in terms of the volatility of returns which leads to the rather confusing notion of negative and positive risk. In business, concepts such as risk aversion and risk appetite are commonplace. This is unfamiliar territory in CRM and not concepts that will be considered further in this book.

This differentiation becomes particularly important when dealing with an adverse incident which has implications in the broader context of risk. For example, suppose a HIT system fails to correctly manage a schedule of patient appointments. This could have safety implications if, for example, patients failed to attend urgent appointments as a result. Likewise;

- Incorrectly reporting on service utilisation could have serious financial implications
- Business risk could be introduced as staff lose productivity from having to manually correcting errors
- Reputational risk could be suffered if the failure to manage resources appropriately was publicised in the press

For those tasked with evaluating clinical risk alone, consequences unrelated to safety need to be consciously removed from the equation and not bias the analysis. As we will go on to discuss, objectivity is paramount in CRM and a tight definition of what constitutes clinical risk versus other types of risk is essential.

2.3 The Perception of Risk

The way we react to risk is determined by the way we perceive it, indeed we have a fundamental need not only to be safe but also to feel safe. Our day to day evaluation of risk is often less than objective and many personal and sociological factors may come into play influenced by our cultural, educational and social-economical background. We may choose to turn a blind eye to some more obvious risk mitigation strategies whilst placing inappropriate confidence in others. In some cases, we may choose to ignore some hazards altogether and place our trust in luck or even a higher authority. Similarly, the perceived benefits of our experiences will vary widely between individuals; the excitement of a parachute jump may be exhilarating for one person whilst terrifying for another. How could one therefore expect those individuals to objectively agree on the relative risk of the same experience?

A significant factor in the evaluation of personal risk is the extent to which one has control over the precise circumstances, particularly whether engagement in an activity is voluntary or enforced, necessary or accidental [6]. Our right to choose whether or not we do something has a direct impact on how we perceive the risk associated with it and, in turn, its acceptability. Similarly we may regard a hazard which we consider to be a necessary evil differently compared with one arising as a result of incompetence or maliciousness. This of course has serious implications for healthcare provision.

Consider the following scenario, a patient is about to undertake a major surgical procedure. The procedure is associated with a significant risk of bleeding, infection and anaesthetic hazards. However the patient, having taken into account the potential benefits of the procedure, consents for it to go ahead. He/she is aware of the facts, is involved in the decision making process and feels assured that due attention is being paid to taking all risk factors into account. Now suppose that procedure is carried out by a trained but relatively inexperienced surgeon when other more qualified members of staff are available. The added risk this brings is probably minor compared to the overall risk of the procedure but, for the patient, the risk-benefit swingometer has made a significant shift. An inability to influence the selection of the personnel, their training and the introduction of an apparently unnecessary hazard has altered the perception of risk disproportionately to its actual degree.

Society expects organisations to be increasingly proactive and forward-thinking in their management of risk [7]. Individuals increasingly perceive a responsibility to be kept safe in their working and leisure activities and consider it the domain of others to provide that level of assurance. Paradoxically, this shift is occurring in a society which is increasingly safe. People now live longer than ever before in history and products are safer and more reliable than at any time in the past.

This analysis might lead one to conclude that subjective views of risk are contaminants to its true and objective nature but this is not the complete picture. For activities which affect the public one could argue that people need to be assured to a level which is tolerable to society irrespective of whether the magnitude of risk is based on scientific fact. If safety is freedom from unacceptable risk [8] (see Chap. 3) and what we consider to be acceptable is influenced by the values of society then one has to conclude that safety is itself at least partly defined by perception.

2.4 Perception of Risk in HIT

If we accept that the perception of risk is linked to a number of extraneous factors then how do these elements affect society's view of HIT? In the main it is probably fair to assume that the public gives little regard to the safety characteristics of technology in healthcare. There is an implicit assumption that the tools we use to support care are fit for purpose, well maintained, in good working order and are operated by trained individuals. These are indeed reasonable expectations so when defects in those systems or the way they are operated introduces hazards this can be difficult to justify to the patient on whom the risk is ultimately imposed.

Most HIT systems are manufactured by private organisations which, by their very nature, exist for the benefit of investors. This arrangement drives the cutting edge of innovation but, to the patient, notions of financial gain in any way linked to their suffering can be uncomfortable. Should a manufacturer's system adversely impact care it is unsurprising this can raise questions about an organisation's priorities. The benefits of HIT systems may be clear to those who deliver healthcare but less clear to those who ultimately receive it.

When one considers the hazardous nature of healthcare delivery, it is likely that the contribution of risk offered by HIT systems whilst present is probably small compared to the dangers of invasive procedures, infection risk, etc. The perception of that risk may be biased by sociological factors but that should in no way diminish our drive and enthusiasm for mitigating risk. In short, we have a duty to undertake clinical risk management of HIT systems because we have it in our power to do so and the public, quite rightly, expect it of us.

2.5 Risk Management

A recent Technical Report from ISO/TC 215 defines risk management as "the systematic application of management policies, procedures and practices to the tasks of analyzing, evaluating and controlling risk" [9]. Similarly ISO 31000 defines it as "coordinated activities to direct and control an organisation with regard to risk" [5]. These definitions are particularly insightful as at their heart is the notion of organised policies and procedures to achieve the objective. In other words, these definitions steer us away at the outset from an ad-hoc, fire-fighting and reactive approach. Risk management is about a careful, logical and justified method undertaken throughout the lifecycle of a product and executed in a controlled and predictable manner. It comes about not as a convenient byproduct of sound engineering but rather something which is undertaken proactively, deliberately and responsibly.

Early risk management systems focussed on retrospectively identifying and analysing untoward incidents. Risk management in these cases really only began when harm had actually occurred. In the 1960s a number of high risk industries recognised that proactive safety management is a preferred strategy. The approach was supported by the development of 'safety management systems' which combine and coordinate numerous methods such as Failure Modes and Effects Analysis (FMEA) into a systematic framework to address safety. In this way steps are taken to avoid harm in the first place through the application of careful thought and design verification.

To some extent the approach to risk management has to be influenced by the nature of the subject matter under consideration. For steady-state, predictable and well documented systems risk evaluation can be performed through the statistical analysis of acquired data using techniques like maximum likelihood estimations, and analyses of variance and correlations [10]. For new systems where little data is available, where human factors are a significant contributor or where systematic failure modes prevail, more subjective measures such as imagination and careful thought are needed to derive potential sources of harm. HIT systems are typically highly complex and fall into this latter category requiring not mathematical expertise to derive the risk but rather an understanding of the domain, human behaviour and best practice.

Effectively managing risk in healthcare and HIT is a complex task but one with incalculable rewards for both manufacturers and healthcare organisations. For example:

- The development of a higher quality product which commands a greater price in the market and reduced support costs
- The building of a safety record creating confidence in the product and attracting new business
- Development of a safe and reputable brand
- Reduction in the cost of managing adverse incidents, legal fees and damage claims
- The forging and integration of best practice
- Improved staff morale and productivity
- Compliance with standards, legal and commercial requirements
- As a driver for continual business and clinical improvement

Once risk is reduced to acceptable levels one is able to truly leverage a system's benefits and reap the rewards of the organisation's investment. Nevertheless one must not lose sight of the fact that of primary importance is a focus on the patient and the risk of harm as a threat to their future well-being. In a litigious society it is easy instead to concentrate our risk management efforts on avoiding legal challenges and claims for compensation. To make this our primary driver is naïve and cynical – ultimately this becomes a strategy to protect ourselves not our patients.

Neither is risk management a tick-box exercise to satisfy a regulator or manager. It is an invaluable tool which we can implement to actually make a difference, save lives and improve the quality of care. Risk management is not about knee-jerk reactions to tackle whatever today's problems might be only to have a different focus tomorrow. It is a strategic, long-term and carefully executed set of activities to address issues before they arise.

Finally one should not lose sight of the fact that risk management requires not just an understanding of the technology but also an appreciation of the complex socio-technical environment in which it is deployed (see Sect. 5.1). A culture of blame, secrecy and complacency simply doesn't allow effective risk management to flourish. Instead the conditions need to be just right; a setting of mutual learning, transparency, proactive awareness and co-operation. Implementing a HIT system successfully requires healthcare organisations to embrace new ways of doing things. By integrating risk management into those processes risk can be systematically controlled rather than naively relying on clinicians to 'be careful'.

2.6 Hazards

To usefully apply CRM to the safety assessment of HIT systems we need to begin with the concept of the hazard. The term hazard is derived from an Arabic word which was adopted to mean an unlucky roll of the dice. There are many modern definitions of hazard some of them helpful to CRM and others less so. What is common to most definitions is the notion of harm. Depending on the context harm might refer to physical injury, financial loss, environmental impact, or a combination of these. In CRM we are referring specifically to patient harm.

One could of course go on to examine the many definitions of harm. Some descriptions of harm relate well to CRM – physical harm for example (represented by changes in a patient's appearance or reported symptoms) is straight-forward to characterise. Other manifestations such as psychological harm or breaches of an individual's privacy are less clear-cut in terms of their relationship to CRM and this is discussed further in Chap. 10. For the purposes of this book, the term harm will be defined as per ISO/IEC Guide 51:1999, "death, physical injury and/or damage to health or wellbeing of a patient" [8].

A reasonable definition of hazard for the purposes of CRM is also provided in ISO/IEC Guide 51:1999 as a "potential source of harm". One may decide to add "… to a patient" to that definition to clarify that harm to a system's operators or those delivering healthcare is generally excluded. Interestingly there are other definitions of hazard which imply something which is inevitable and cannot in any way be avoided. This clearly flies in the face of risk management and such a definition would presumably preclude the existence of controls. Hazards are therefore entities which may or may not in the real world result in actual harm but certainly have the opportunity to do so should they fail to be adequately controlled.

An important property of the hazard is its impact or consequence (the two words being largely interchangeable). A true hazard will have a credible and clearly apparent effect on one or more patients. In other words it should be possible to draw a line of cause and effect from the hazard to patient harm. Were it is not possible to draw this line, the existence of the hazard or the manner in which it had been described may need to be revisited.

A review of the potential consequences of a hazard should allow an individual with domain knowledge to take a view on the seriousness of the hazard's impact. For example, would the hazard typically lead to consequences that would precipitate a patient's demise or merely introduce a degree of inconvenience? That assessment will lead to the development and definition of the severity of the hazard. Note that despite frequent confusion in the literature, severity is not the same as impact or consequence. Severity represents the degree of harm whilst impact or consequence defines the nature of that harm. In other words it is examination of the impacts and consequences which allows us to estimate the degree of severity.

Table 2.1 shows some examples of hazards along with potential impacts. For simplicity only one impact is shown but often there are multiple impacts for a single hazard.

2.7 Components of Risk

To manage risk effectively it is necessary to undertake a deep dive to understand its constituent components and how those factors interrelate.

Table 2.1 Examples of hazards and their potential impact

Hazard	Potential impact
A prescriber could select the wrong drug to be administered to a patient	The patient could receive an inappropriate medication whilst failing to receive the correct medication
The system could fail to display the required care pathway flowchart for a particular condition	A clinician reliant on the flowchart to inform their clinical decision making could provide inappropriate care
Laboratory results could be allocated to the wrong patient	A clinician could make clinical decisions on information which pertained to the wrong patient whilst the target patient failed to receive their results
Documents sent to the remote network printer could fail to print resulting in a backlog or lost documents	Access to care could be delayed or care providers could fail to receive clinical correspondence which they depend upon

2.7.1 Risk and Probability

So what precisely is risk? One might propose that risk is simply the probability of something going wrong. Indeed in everyday language we use terms like probability, chance, likelihood and risk interchangeably. This use of the term is however overly simplistic when it comes to formal risk assessment and evaluation. Take the following (theoretical) scenarios:

Study A demonstrates that one in a thousand patients is regularly administered double the required dose of Insulin.

Study B demonstrates that one in a thousand patients is regularly administered double the required dose of Paracetamol (Acetaminophen).

The question is, is the risk to the patient the same in both these scenarios? If we take the definition of risk as merely an analogue for probability, the risk is presumably the same – the likelihood of receiving double the recommended dose is 1 in 1,000 in both circumstances. Somehow that feels wrong. The narrow therapeutic window for Insulin compared to Paracetamol means that harm would be far more likely to occur in the case of Insulin. Thus, the likelihood is the same but the severity of the outcome is different.

Note: One could generate many similar scenarios where the outcomes were very different. For example, if you were dealing with a patient with significantly impaired liver function, administering double the dose of Paracetamol could be significant. As we will go on to discuss later, it is the typical picture that is important rather than attempting to define the exceptions and all possible scenarios.

We can therefore see that the concept of risk begins to emerge when we combine the likelihood of a hazard occurring with the severity of its impact. It is therefore proposed that a potential definition of clinical risk in the context of CRM could be "the chance of a patient being harmed to a stipulated extent"; the chance component is defined through the hazard's likelihood whilst the stipulated extent is defined by the severity. In other words:

$$\text{Risk} = \text{Severity} \times \text{Likelihood}$$

This definition also fits neatly with that in ISO 14971 [11], "combination of the probability of occurrence of harm and the severity of that harm". Note that whilst risk in this case is defined in terms of the combination of severity and likelihood interpreting this as a formal mathematical product can be problematic. Suppose one was to assign a score to each variable, multiply them together and use the result as a measure of risk. It quickly becomes apparent that the formula provides a non-linear result with risks at the higher end of the spectrum having a disproportionately inflated score. Similarly a rather curious situation emerges in that it would be perfectly reasonable to derive a risk with a score of 16 or 18 but never 17 – or any other prime number for that matter. The difficulty arises from the multiplication of ordinal numbers which is a by no means a straight-forward subject. Instead there is a strong case for employing a simpler methodology based on the categorical combination of severity and likelihood.

In defining clinical risk it should also be pointed out that the term varies its meaning depending on whether the definite or indefinite article precedes it. The definition presented above holds true when one talks about 'the clinical risk' but from time to time we use the phrase 'a clinical risk' to describe something which could give rise to harm. For example, 'We believe that the incorrect information on this screen represents a clinical risk'. In most cases a more correct term would be hazard (i.e. 'a potential source of harm'). Nevertheless 'clinical risk' is used so frequently in this way that to deem it incorrect invariably elicits an accusation of pedantry. Still, hazard is a preferable term in formal text to avoid any ambiguity especially when faced with awkward phrases such as 'the clinical risk of the clinical risk is low'.

2.8 Quantifying Risk

Once we have the formula and definition of risk, one needs to consider how that can be used in the context of an SMS to evaluate risk. To allow us to combine the components of risk we need to develop:

1. A series of severity categories
2. A series of likelihood categories
3. A series of clinical risk categories
4. A matrix indicating which clinical risk category applies to each combination of severity and likelihood (Table 2.2).

The resulting matrix is a simple way to communicate and evaluate risk in a repeatable and consistent manner. At the point of evaluating risk each unique severity category can be paired with a unique likelihood category and this combination points to a non-unique clinical risk category. Once embedded, this simple grid acts

Table 2.2 Example risk matrix

		Severity				
		Minimal	Minor	Moderate	Severe	Catastrophic
Likelihood	Regularly	Medium	Medium	High	High	High
	Often	Low	Medium	Medium	High	High
	Occasionally	Low	Low	Medium	Medium	High
	Infrequently	Very low	Low	Low	Medium	High
	Rarely	Very low	Very low	Low	Low	Medium

as a cornerstone (or rather, an organisation's cornerstone) of risk evaluation. Many organisations may already be equipped with a risk matrix and, assuming the categories are flexible enough, this could be harnessed for CRM purposes. Not only does this save re-work but also demonstrates the integration of CRM activities with a wider organisational risk management strategy.

Note that some safety critical industries operate with standards which mandate the use of a particular risk matrix. Currently this is not the case for HIT in most countries (although some have been voluntarily adopted from non-HIT patient safety programmes [12]). Organisations are free to create a matrix which works well for them – the price paid for this flexibility is a lack of consistency. This can be a particular challenge when a HIT system manufacturer is using a different risk matrix to that operated by the healthcare organisations they supply. This is a situation which adds to the complexity of managing risk across organisational boundaries especially when risk appetite varies.

2.8.1 Developing Risk Matrix Categories

There are a couple of approaches that can be taken in developing working categories for severity, likelihood and clinical risk. Clearly it is important to make each category distinct, discrete and with an easily understood sense of order. A numerical or quantitative method is one option. For example, likelihood could be classified as an incident which occurs; once per day, once per month, once per year, etc. A common alternative is to use word pictures to define each category, a qualitative approach. For example, for likelihood this might be; Rare, Occasional, Often, Always or for severity; Minor, Major, Catastrophic, etc. Usually the word is supplemented with a qualifying description to clarify the relative order of each category. For example; Rare – hardly ever occurs, Occasional – happens from time-to-time, Often – occurs regularly. Ultimately the category descriptions should be made relevant to the subject matter. The more vivid a story is painted the more likely stakeholders will agree on a category. Examples of suitable categories exist in the academic literature and standards in this area (such as ISO/TS 25238:2007 [13]).

One might argue that such a simplistic, qualitative categorisation of severity and likelihood is open to interpretation and is subjective. Surely the more objective approach would be quantitative based on concrete probabilities and mortality/morbidity rates? This may be true, but the qualitative methodology is associated with some complexities in an HIT CRM setting.

The difficulty comes in determining how frequently HIT systems, and indeed individual functions and behaviours are responsible for patient harm. Research in this area is very limited indeed and, in the main, the data simply does not exist. Any attempt to apply a quantitative analysis to the failure of an HIT system quickly becomes estimation and guesswork. One can quickly become trapped in the belief that because the methodology has the illusion of objectivity then the outcome must be valid irrespective of the underlying data. A numerical approach immediately opens the debate of ‘What is your evidence?’ Ultimately the methodology results in a situation that is just as subjective as using simple qualitative categories for severity and likelihood but can be more time-consuming and complex.

So how could risk categorisation ever work in such a sea of subjectivity? We need to go back to the primary reasons for undertaking a CRM analysis. In any CRM project the reality is that risk reduction activities require resources and those resources are finite. If risk reduction activities are spread evenly across a system some areas, perhaps associated with significant risk, will fail to get due attention. Similarly, less risky areas will be the subject of undue focus. Much of CRM is about prioritisation, reducing the risk in those parts of the system that are known to be the greatest contributors. If one compares one hazard with another, i.e. the ratio of two clinical risks, any units of measure or other factors are cancelled out in the analysis rendering any absolute measures redundant.

Once it is accepted that determining clinical risk is largely about prioritisation, one can see that the estimation of risk can be relative rather than absolute. Essentially it is more useful to know that ‘Functionality A’ contributes significantly more risk than ‘Functionality B’ than it is to know that ‘Functionality C’ is estimated (with absolutely no supporting evidence) to contribute to 1 in 10,000 deaths. The message is that so long as the meaning of each category is clear and those definitions are applied correctly and consistently by the individuals evaluating risk, then a qualitative approach provides a practical means of focussing CRM effort and risk reduction activities in most cases.

As many risk evaluation techniques could be seen to have subjective elements it is essential that measures are taken to carefully manage any subjectivity during risk evaluation. These measures might make us of:

- Domain expertise – There are many complex factors to take into consideration when establishing severity and likelihood. These need to be examined by individuals with experience in the clinical domain.
- Awareness of bias – Bias is discussed in Sect. 15.2. Given that the selection of a severity and likelihood category involves a judgement call, there is an opportunity for bias to contaminate the evaluation. Should an organisation have an agenda for the clinical risk evaluation to have a particular outcome, it can be tempting to express a hazards in terms of a special case. It is therefore essential

to remain objective at all times and to avoid using clinical risk as a bargaining tool for an individual to get their own way.

- Consensus – Both the establishment of domain expertise and the elimination of bias can be facilitated by obtaining a consensus view of the severity and likelihood. The optimal balance is found when stakeholders across organisational and departmental boundaries are able to meet and collectively agree the evaluated clinical risk.

We should now consider what precisely is meant by severity and likelihood in the context of HIT.

2.8.2 Severity

Severity represents the magnitude of the impact or potential impact on the patient – in other words, the extent to which a patient's wellbeing could be compromised. Note that the key word here is 'patient'. One should only begin to determine the severity of a hazard when the potential impact on the patient can be and has been ascertained. If it cannot be established, further investigation is usually necessary (in association with a domain expert) or consideration should be given as to whether the scenario is indeed a hazard in the first place.

Putting the patient at the centre of the analysis forces one to consider the impact on the patient rather than that on the system or its users. For example:

System A is used to schedule screening tests for adult patients. A hardware fault occurs which results in system downtime for a period of 6 h.

System B is used to determine what medication is due to be administered to in-patients. A hardware fault occurs which results in system downtime for a period of 2 h.

So how does the severity differ between these two scenarios? From a purely engineering perspective, one might argue that the outage of System A is three times more severe than that of system B – i.e. 6 h compared to two. But this simplistic approach considers only the impact on the system not on the patient. An individual with domain knowledge will know that screening tests are rarely scheduled on an urgent basis and, whilst an operator of System A may have a backlog to work through, ultimately they will probably catch up. On the other hand, unavailability of System B could immediately result in one or many patients failing to have medication administered. A delay of up to 2 h could be clinically significant for some short half-life medications. Thus, from a patient perspective, the issue with System B will result in an outcome of higher severity.

Given that a particular fault in a system could ultimately give rise to a wide variety of potential and largely unpredictable consequences, how could anyone ever go about objectively evaluating the severity? For example, imagine a system fails to schedule clinic appointments correctly. What severity would one choose for the impact? One appointment might be a routine follow-up for a minor injury whilst another could be in relation to a serious, recently identified malignancy requiring urgent treatment. Surely the only way to manage this would be to assign a clinical risk to each and every patient? Thankfully this is not required.

To rationalise this dilemma, it is necessary to consider the 'typical' or 'most frequent' outcome and assess the severity of that event. In other words, this is an educated judgement call based on previous experience in the clinical domain. In this case one would need to consider, across all specialties, appointments types, consultants, medical conditions and patients, what is the typical and most appropriate severity category associated with an appointment not being scheduled? Whilst this might not stand up to the detailed scrutiny of the scientific method, remember that what is important is the relative magnitude of the severity compared with other hazards under consideration.

Note that in debating the clinical risk associated with a hazard in a group it is all too tempting to propose special cases that describe scenarios with severities in all categories across the entire severity spectrum. Despite this divergent thinking, stakeholders will probably agree quite quickly that the severity of the target hazard is greater or less than that of another. It is this relative and convergent view that is important and allows us to abstract the analysis to a workable level.

2.8.3 *Likelihood*

Compared to severity, likelihood can be a surprisingly ethereal concept. At the simplest level one can equate likelihood with probability or chance – essentially a statistical interpretation of whether an event may or may not actually happen. The complexity arises when one attempts to define which event is actually under examination.

The path from a hazard being triggered to actual harm occurring is not straightforward. Variables in the way that clinical information is used, how the system is implemented and the actions taken by users in response to system behaviour all come into play. At any stage harm may be prevented by human intervention, a so-called 'learned intermediary' between the technology and the patient. ISO 14971 Guidance [14] recognises this and splits out the concepts of Harm and a Hazardous Situation acknowledging that typically a Hazardous Situation arises first and may or may not lead to Harm. Technically one is required to examine the likelihood of each stage of the path to harm and combine these to derive the overall likelihood.

Whilst it is quickly possible to get bogged down in mathematical detail, the take-away message is to ensure that in the estimation of likelihood one looks at the probability of actual harm occurring. Stopping elsewhere along the hazard-to-harm path is fraught with difficulty as the estimated likelihood will depend on exactly which event constitutes harm in the eyes of the assessor. Similarly, by looking at the end-to-end chain of events the presence or absence of controls can be factored in to the likelihood estimation.

The situation is complicated further when one begins to combine likelihood and severity to estimate risk. This complexity is not just of academic interest but represents a very real challenge which is capable of plaguing risk evaluation. Consider the following example:

What is the clinical risk associated with rose pruning? An overenthusiastic or careless gardener could accidentally suffer a puncture wound from a rose thorn. This might happen regularly for a keen gardener. Potentially an individual could develop an infection in the wound and even septicaemia as a result. This could lead to death, a catastrophic event. When this is put into a risk matrix, i.e. a frequent likelihood and catastrophic severity, it is likely that the result will be evaluated as high risk.

On this basis given the predicted tide of amateur horticulturalists barely clinging on to their very survival, surely gardening should be a regulated activity conducted only by trained professionals under careful conditions? This scenario feels absurd and yet, it is a thought process that is often followed in many a CRM workshop. The errors in logic here are twofold:

1. Firstly the individual proposing this analysis has fallen into a trap. They have taken the worst case scenario, the special case – septicaemia. The typical outcome of a thorn puncture would not be death but perhaps mild pain and some concentrated effort with a pair of tweezers.
2. Secondly, and key to our likelihood discussion, the proposer has asked the question, 'What is the likelihood of puncturing your finger with a rose thorn?' The question he/she should have asked is 'What is the likelihood of puncturing your finger with a rose thorn AND this leading to an event of catastrophic severity?' This would yield a very different result.

This example leaves us with an important message which, if followed, will avoid us falling into this trap:

1. Select the severity first – use the typical severity based on experience in the real world
2. Select the likelihood in terms of experiencing an event of the chosen severity and not of the cause being triggered

Another strategy in circumventing this issue is to replace the term 'likelihood' with 'likelihood of harm'.

Thus, having clearly flagged the tempting but ultimately misguided path, we can finally define likelihood as being the probability of experiencing an adverse event of the defined severity.

Interestingly guidance for Standard ISO 14971 [14] provides a slightly different view on the estimation of likelihood. Here it is acknowledged that the probability of software failure is difficult to predict and allows for the assessor to instead estimate risk on the basis of the severity of the harm alone – in other words, a failure probability of 1. This approach clearly relieves some of the difficulties in this area. However one immediately loses the ability to recognise that, relative to each other, some failure modes are clearly more likely than others. Even if an accurate, evidenced-based calculation of probability is troublesome, expending effort in deriving at least some order of magnitude for likelihood has significant merit when it comes to prioritising risk reduction measures.

2.8.4 *Risk Equivalence*

In 2.7.2 we discussed the difficulty one is faced with when a group of individuals have different views on the level of severity for a particular hazard. Indeed the same problem exists in evaluating likelihood – so how can one ever come to a consensus when managing a number of stakeholders with different opinions? The solution often comes from working through the evaluation to its conclusion.

Inspection of the risk matrix reveals a series of diagonal stripes within which the clinical risk remains the same. Different combinations of severity and likelihood typically result in the same clinical risk. If individuals disagree on the components of risk they may well agree on the derived clinical risk. One person's Moderate-Remote might be another person's Minor-Occasional ultimately both concluding on the same degree of risk.

Note that in a group setting when a consensus is being sought it often a good idea to propose a severity, obtain agreement and then discuss the most appropriate likelihood. This avoids both the gardener's mortality trap described above as well as a barrage of differing yet pointlessly equivalent severity-likelihood combinations.

2.9 Summary

- Risk is an inevitable part of life and presents itself in many different forms.
- The way we perceive risk is affected by more than just its absolute magnitude.
- Society is less tolerant of unnecessary risk and patients almost certainly consider risk from HIT to be unnecessary.
- Risk management is undertaken in an attempt to reduce risk. Proactive approaches are preferable to reactive methods.
- A hazard, in the context of CRM, is a potential source of harm to a patient and a level of clinical risk can be attributed to a hazard.
- Clinical risk can be derived by examining the severity of a hazard's impact and the likelihood of experiencing an event of the stimulated severity.

References

1. Lowrance W. Of acceptable risk: science and the determination of safety. Los Altos: W. Kaufman; 1976.
2. National Advisory Group on the Safety of Patients in England. A promise to learn – a commitment to act. Improving the safety of patients in England. Br Med J. 2013. https://www.gov.uk/government/uploads/system/uploads/attachment_data/file/226703/Berwick_Report.pdf.
3. Wald H, Shojania K. Making health care safer: a critical analysis of patient safety practices. July 2001.
4. Oxford English Dictionary. [Online]; 2015. Available from: http://www.oed.com/.
5. International Organization for Standardization. ISO 31000:2009, Risk management – Principles and guidelines. 2009.

6. Fischoff B, Slovic P, Lichtenstein S. How safe is safe enough? A psychometric study of attitudes towards technological risks and benefits. Policy Sci. 1978;9:127–52.
7. Health and Safety Executive. Reducing risks, protecting people. HSE's decision making process. 2001.
8. International Organization for Standardization. ISO/IEC Guide 51:1999. 1999.
9. International Organization for Standardization. ISO/TC 215 N1143 – ISO/DTR 17791 – Health informatics — guidance on standards for enabling safety in health software. 2013.
10. Paté-Cornell M. The engineering risk analysis method and some applications. Chapter 16. [Online]. [cited June 2015; http://www.usc.edu/dept/create/assets/002/50856.pdf]. Available from: http://www.usc.edu/dept/create/assets/002/50856.pdf.
11. International Organization for Standardization. EN ISO 14971:2012. Medical devices. Application of risk management to medical devices. 2012.
12. UK National Patient Safety Agency. A risk matrix for risk managers. London: UK National Patient Safety Agency; 2008.
13. International Organization for Standardization. ISO/TS 25238:2007 (Health informatics – classification of safety risks from health software). Technical Specification. 2007.
14. International Electrotechnical Commission. IEC/TR 80002-1:2009. Medical device software, Part 1, Guidance on the application of ISO 14971 to medical device software. 2009.

Chapter 3
Acceptability and Ownership of Risk

3.1 How Safe Is Safe?

Ultimately this is perhaps the most difficult question to answer. This dilemma of risk acceptability is more than just academic, in fact it goes to the very heart of our definition of safety; safety is freedom from unacceptable risk [1]. So how does one go about establishing to what extent risk is acceptable?

For one, it depends on your point of reference. As a patient undergoing a surgical procedure one would presumably choose to receive care of the utmost quality; a surgeon and anaesthetist who were the best in their field using state of the art equipment and perfectly maintained facilities. The reality is that the quality of care we provide and receive exists somewhere on a spectrum. We have to reach a balance between quality, available skills and resources which is acceptable to all parties and to us as a society.

So how does this dilemma translate into HIT and CRM? Suppose an organisation knows that an HIT system has a particular defect which introduces a degree of clinical risk. Are the stakeholders involved not duty bound to fix the issue immediately whatever the risk? If a patient was to die or experience serious harm as a direct or indirect consequence of the issue would the manufacturer and/or healthcare organisation be immediately negligent? To take this to its logical conclusion, any risk associated with an HIT solution is surely intolerable.

Over time the more frequent use of the term 'acceptable risk' in standards and guidelines is notable [2]. For example, the International Organization for Standardization report 'Safety aspects: Guidelines for their inclusion in standards' [1] defines Tolerable Risk as being "risk which is accepted in a given context based on the current values of society". Similarly the UN defines acceptable risk as "the level of potential losses that a society or community considers acceptable given existing social, economic, political, cultural, technical and environmental conditions" [3]. Of course society shows their level of risk acceptability by foot-voting – if there was a plane crash every week no one would fly. Ultimately acceptable risk

A. Stavert-Dobson, *Health Information Systems: Managing Clinical Risk*,
Health Informatics, DOI 10.1007/978-3-319-26612-1_3

cannot be framed in terms of an absolute measure but rather by what we as a society consider reasonable.

The debate about what is acceptable risk has existed outside of HIT for decades and was famously tested in the UK court of appeal in 1949 [4]. The case related to an industrial accident rather than healthcare delivery but the ruling by Lord Asquith of Bishopstone had far reaching consequences across many safety-critical industries. The judgement concluded that it was appropriate to consider the sacrifice in money, time or trouble in the measures necessary for averting an incident.

On this basis the concept of 'so far as is reasonably practicable' (SFAIRP) was born and subsequently enshrined in the 1974 UK Health and Safety at Work Act. SFAIRP, whilst a significant step forward, fails however to precisely acknowledge the notion of risk and its relationship to practicability. As such the slightly modified term 'as low as reasonably practicable' (ALARP) has been widely adopted in the UK and some other countries as the basis for risk acceptability. The ALARP principle can be defined as "that level of risk which can be further lowered only by an increment in resource expenditure that is disproportionate in relation to the resulting decrement of risk" [2].

The notion of ALARP is simple yet fundamental. Firstly it acknowledges that there is a difference between clinical risk being 'as low as possible' and what can realistically be achieved. This important principle brings acceptance that organisations do not have an infinite amount of money, time and effort and that there are practical and technical boundaries on how far, from a safety perspective, one can go to reduce risk. In fact one could argue that to continue to reduce risk beyond what is reasonably practicable becomes an inappropriate use of resources and ultimately of little benefit.

Secondly, the idea takes into account the application of proportionality – a match between the resourcing of risk reduction activities and the degree of risk itself. One has the freedom to segment a product or service and apportion risk reduction effort to where it is needed the most. Importantly the concept allows us to develop a considered argument to justify that distribution of effort and potentially tackle any criticism of a failure to apply risk management to an equal extent across a product or service.

Finally ALARP introduces a basis for justifying risks, to make a case for them being present but tolerable. To justify something in this way requires careful analysis and logic set out as an evidence-based argument. The vehicle for this justification is a document known as a safety case.

The ALARP approach promotes flexibility and self-regulation with some sturdy case law as a base. However the approach is not without its critics [5]. There is no formula for ALARP and its open-ended nature begs the question of what really is reasonable and practicable. Similarly evidencing completeness, that all things which can practically be done have been done, can at times be somewhat subjective and problematic. Despite ALARP being the basis for risk acceptability in standards like ISB 0129/0160 [6, 7], at the time of writing ALARP remains largely untested.

Critics of ALARP often subscribe to the notion of minimum risk (rather than what is practicable) but this too has its shortcomings. Even if one accepts that risk decreases down a curve as we invest in risk reduction activities, it is unlikely that that curve ever reaches zero but rather goes on to infinity. Logically we must assume that to achieve minimum risk we need to invest an infinite amount of money and effort – something that is simply not realistic. Thus whilst non-ALARP objectives might represent a safety utopia, by defining that as our practical goal we are instantly setting ourselves up to fail.

3.2 Acceptability Stratification

IEC 61508 [8] describes how the clinical risk associated with each identified hazard can broadly be classified into three distinct categories. This is illustrated by a diagram popularised by the UK Health and Safety Executive [9] which has become known as a carrot diagram (Fig. 3.1). The vertical axis represents the degree of risk whilst the width of the carrot signifies the amount of risk reduction effort one must be able to justify to claim that further mitigation is not practicable.

At the lower end of risk there is acknowledgement that most systems will be associated with some hazards considered to contribute minimal clinical risk. In HIT these background risks are often intrinsic to the clinical environment in which the system is operated. In this region it can usually be justified that the small amount of risk is simply not worthy of additional mitigation. Indeed, to focus risk reduction resources in this area would typically not be fruitful given the benefit gained. These

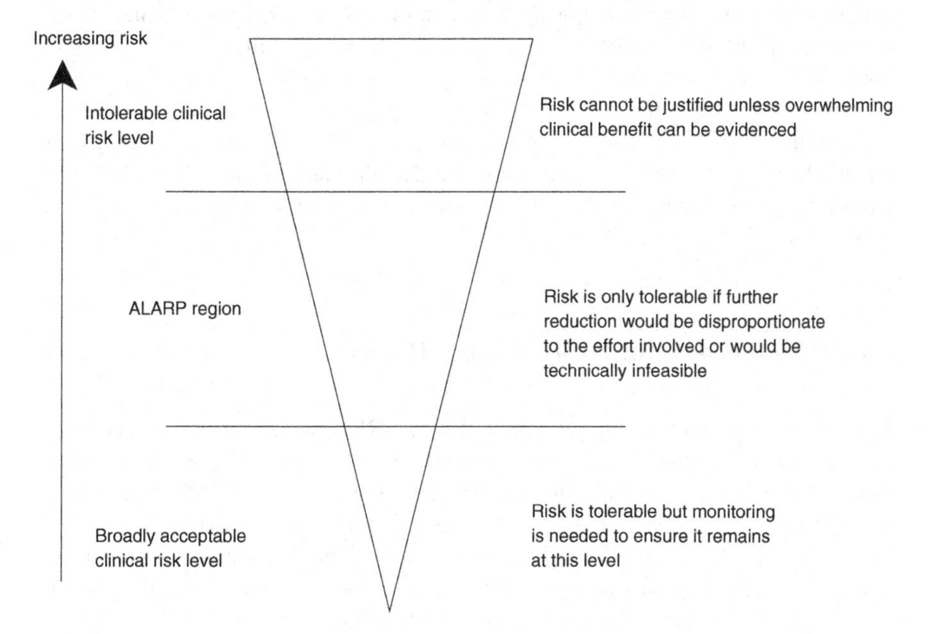

Fig. 3.1 Carrot diagram

risks are therefore generally considered acceptable. They are however not ignored. The nature of risks can change during the lifecycle of a product and it is necessary to routinely revisit hazards in response to change so that one can confirm the fact that they remain in the acceptable range.

At the other extreme, there are some hazards which are associated with so much clinical risk that they could never be tolerated (unless significant clinical benefit could be exceptionally evidenced, see Sect. 3.5). For example, suppose a healthcare organisation plans to implement an electronic prescribing system which, for technical reasons, requires 8 hours of downtime between midnight and 8 am each day. Clearly the complete lack of any ability to manage medications through the night poses a very significant risk indeed. Risks at this level are generally considered to be unacceptable and should they be identified it will typically be necessary to revaluate the project, perhaps consider re-engineering or a modification of the project scope. Often this occurs long before a safety case is attempted.

Finally there are those risks that fall in the middle, frequently the home for many of the hazards found in HIT. In this region, a new factor comes into play. Basically a risk is considered tolerable providing it has been mitigated to as low as reasonably practicable. In other words it is not good enough to simply identify the hazard, allocate it to the risk middle-ground and forget about it. To do so would be to define a risk that is unacceptable or, at best, unjustified. As long as one can, through careful consideration and balanced argument, evidence that reasonable and proportionate measures have been put in place to mitigate the risk then tolerable clinical risk can be demonstrated. Note though that the higher the initial risk the greater the burden of disproportionality becomes. By tolerable we imply (and indeed assume) that our patients would be willing to live with the risk in the knowledge that it is being appropriately mitigated within practicable limits. Thus the acceptability of clinical risk in this region is not a function of the absolute degree of risk but rather the extent to which risk has been mitigated through the reasonable application of effort and resources.

Note that there is no real concept of 'temporary ALARP', one cannot accept a non-ALARP solution whilst a permanent ALARP alternative is sought. A particular control may offer a temporary fix but the resulting risk is either acceptable or not at a given point in time.

3.3 Risk Acceptability of Medical Devices

At this point it is worth mentioning how the ALARP approach to risk acceptability changes markedly when it comes to medical devices. For example, International Standard ISO 14971 [10] affords manufacturers the ability to justify risks on an ALARP basis. The European Commission Medical Device Directive (MDD) [11] cites ISO 14971 as a Harmonised Standard; however the directive contains a number of Essential Requirements on risk acceptability which appear to conflict with ISO 14971.

As such, the European National version of the Standard (EN ISO 14971:2012) via its Annex Z sets the risk acceptability bar differently, "All risks, regardless of their dimension, need to be reduced as much as possible and need to be balanced, together with all other risks, against the benefit of the device." In other words one cannot make an argument to say that additional risk reduction is not required as the effort and cost would be disproportionate to the mitigation achieved. Instead all possible controls must be implemented unless:

- The control would not actually reduce the risk, or
- There is a more effective risk control that cannot be simultaneously implemented

Note that even risks which are considered to be low and might otherwise be taken as acceptable by default still require mitigation if this is at all possible. The seemingly bottomless pit of risk mitigation can be qualified by a risk benefit analysis, i.e. the ability to justify the risk against the product's benefits rather than the cost or effort of further risk reduction. The risk benefit analysis also needs to include the benefits of using an alternative product.

The MDD was developed primarily to regulate physical products with a tightly defined intended purpose. In these cases it is relatively straight-forward to undertake clinical trials and literature reviews to construct the evidence underpinning the risk benefit argument. In contrast, developing the same level of data for complex and highly configurable HIT systems can be extremely challenging. Many HIT systems sit outside the highly regulated medical device environment and benefit from the practicable freedom which ALARP affords. For those which cross the divide the strategy is often one of tightly defining the precise components which are subject to regulation and have an intended purpose which is narrow enough to facilitate data gathering.

3.4 Who Owns the Risk?

The risk which remains after controls have been put in place is called residual risk (as opposed to initial risk) and the question arises as to who 'owns' this risk and specifically whether it is a shared or apportioned entity.

A great deal of time and effort can be expended in a CRM project discussing which party owns the residual risk. This conversation usually occurs across organisational and commercial boundaries often between suppliers and their healthcare organisation customers. Unfortunately this debate can be a cynical one and can quickly descend into an attempt to prophylactically apportion blame. It is a common misconception that the purpose of the safety case is to 'transfer' risk onto other stakeholders such that when an incident occurs one is legally able to fall back on the safety case to demonstrate that all risk and therefore liability has been 'transferred away' from the authoring party. Instead the safety case is there to acknowledge hazards and agree mutual and workable strategies for risk reduction.

Importantly there is a distinction between liability, risk acceptability and ownership. Liability is a complex area and determined by the nuances of commercial contracts and the law of the land. Ultimately liability is decided upon by a court and may or may not involve the risk assessment. Acceptability relates to the decision of whether a product can go to market or go live based on its risk profile. Any organisation who makes an acceptability evaluation essentially owns that decision. Another view is that the only party who can truly accept residual clinical risk is the individual who is the subject of harm – i.e. the patient. How one could make this acceptance work in practice is a different matter of course but the notion is important in principle. Perhaps a more practical interpretation is that the healthcare organisation, in making a decision to implement and go live with a system, is accepting risk on the patient's behalf (as well as on behalf of itself).

Finally, risk ownership is about who takes practical responsibility for ensuring that risk is managed, that controls are put in place and validated accordingly. A well authored control will have a clear indicator of which stakeholder will be responsible for its implementation and for evidencing its effectiveness. Should a control be found to have been ineffectively implemented, a collective and objective discussion can be had regarding possible options and making use of commercial arrangements to effect a change where necessary. In this way the management of risk can be shared and be the subject of collaboration without the need to artificially and emotively apportion blame.

3.5 Risk Versus Benefits

Ultimately one could judge the acceptability of risk on a very simple criterion. Do the benefits of the proposed measure outweigh the risks? Indeed Kaplan and Garrick argue that in isolation there is no such thing as acceptable risk; that because of its very nature risk should always be rejected, and it is only the chance of benefits that leads us to tolerate certain risks [12]. On a day to day basis we manage the hazards of everyday life on a risk benefit argument. We know that driving a car is a risky activity but the benefit a personal vehicle brings largely outweighs that risk. Could we not use that same argument as a crude means of safety assessing an HIT system? In other words, are we justified in subtracting a product's risk from its benefits, demonstrating a positive result, and use this as the basis for the safety case?

This is one area, like many others in HIT, where the devil is in the detail. For simple and well defined software applications, this argument may be reasonably straight-forward to construct. For example, the software embedded within an implantable cardiac pacemaker is integral to its operation and makes functional a device which has overwhelming and live-saving benefits to patients. Those benefits are straight-forward to derive and, with some analysis, can potentially be quantified.

However, when one considers more complex HIT applications such as the electronic patient record or order entry systems the risk-benefit argument becomes less straight-forward. To begin with, the benefits of HIT systems are often diverse and challenging to measure. A typical HIT system might deliver cost-savings, improved efficiency, convenience and of course improvements in patient safety. But how would one go about objectively measuring and comparing those benefits with the clinical risk associated with introducing the product? Any judgement would likely be highly subjective. And who would make that judgement? The finance director, clinician, patient and indeed the tax payer may well have very different views.

Secondly, the risk-benefit argument becomes confused if the individual enjoying a benefit is different to the individual being exposed to risk. With HIT systems, it is ultimately the patient who is the subject of harm and therefore risk yet the system may provide enormous benefit to the finance director. To apply the risk-benefit argument one would not only need to be able to measure benefits but also to frame these from the patient's perspective. For most HIT systems this would indeed be a complex analysis.

Thirdly, if one is not careful, a crude risk benefit analysis can lead to gross over simplification – almost to the point of justifying a product's risk in nothing more than a few sentences. When one is faced with the effort and expense of undertaking a careful CRM assessment it might on occasion be tempting to take the following position:

1. This software is used to treat patients who are critically ill.
2. If we fail to treat patients who are critically ill they will surely die.
3. Death is the worst possible outcome.
4. The implementation of the software is therefore justified.

…there we have it, a safety case in four lines – albeit one that is useless and fulfils no purpose whatsoever other than to tick a box. If one decides to go down the line of a risk-benefit approach, this should in no way detract from the degree of rigour required to assure the system.

A number of approaches such as those set out in ISO 14971 [10] and ISB 0129/0160 [6, 7] take the view that benefits should only be taken into consideration when the residual risk has been determined to be otherwise unacceptable and that further risk control is not practicable. Note that there is clearly no guarantee that the risk will become acceptable once benefits have been considered and any analysis should certainly not begin with this assumption.

Any risk-benefit analysis should be clearly set out in the safety case with the appropriate justification, rationale and supporting evidence. The complexity of the arguments involved warrant close cooperation between manufacturers, users and regulators. In most cases the inclusion of potential benefits in the safety case is simply not required in HIT CRM and, if anything, introduces subjective noise into an otherwise objective methodology.

3.6 Summary

- The extent to which risk is acceptable can be defined in a number of different ways depending on the frame of reference.
- The notion of reducing risk to 'as low as reasonably practicable' or ALARP has been widely adopted in some countries. This measure creates the necessarily flexibility to allow the justification of residual risk using a safety case.

- CRM is not a vehicle for apportioning blame. It can be used however to communicate controls, affirm their implementation and justify the acceptability of any residual risks.
- In HIT there are practical difficulties using a risk / benefit analysis as the basis for a safety argument.

References

1. International Organization for Standardization. ISO/IEC Guide 51:1999. 1999.
2. Manuele F. Acceptable risk – time for SH&E professionals to adopt the concept. Professional safety. 2010.
3. The United Nations Office for Disaster Risk Reduction. [Online]. 2015 (cited Mar 2015). Available from: http://www.unisdr.org/we/inform/terminology.
4. Edwards v NCB. 1 KB 704, [1949] 1 ALL ER 743. 1949. http://www.newlawjournal.co.uk/nlj/content/being-reasonable.
5. Bibb B. The medical device risk management standard – an update. Safety-Critical Systems Club – Newsletter. 2005;14(3).
6. Health and Social Care Information Centre. ISB0129. Clinical risk management: its application in the manufacture of health IT systems version 2. UK Information Standard Board for Health and Social Care; 2013.
7. Health and Social Care Information Centre. ISB0160. Clinical risk management: its application in the deployment and use of health IT. UK Information Standard Board for Health and Social Care; 2013.
8. International Electrotechnical Commission. International Standard IEC 61508: functional safety of electrical/electronic/programmable electronic safety related systems. Geneva. 2000.
9. Health and Safety Executive. Reducing risks, protecting people. HSE's decision making process. 2001.
10. International Organization for Standardization. EN ISO 14971:2012. Medical devices. Application of risk management to medical devices. 2012.
11. Council of the European Communities. Council Directive 93/42/EEC of 14 June 1993 concerning medical devices. 1993.
12. Kaplan S, Garrick J. On the quantitative definition of risk. Risk Anal. 1981;1(1):11–27.

Chapter 4

Standards and Quality Management in Health IT

4.1 What Are Standards?

A standard is an agreed way of doing something [1]. A number of organisations publish standards in all manner of subject areas for a local, regional or global audience. They typically encompass a body of knowledge derived from a number of experts in their field with the intention of sharing best practice and forging consistency. The development of standards is market-driven and a number have been formulated to support the creation and implementation of HIT. A small number relate directly or indirectly to the management of risk and patient safety.

The institutions which publish standards do not require organisations to comply specifically. They are created in the interest of mutual learning and sharing of information. However commercial organisations may choose to make those whom they do business with comply through commercial contracts and regulators may enforce compliance in law.

Many standards involve both normative and informative content. Normative information essentially represents requirements, i.e. activities that one must undertake in order to comply with the standard. Informative content on the other hand suggests how one might (but not necessarily) go about implementing the requirement – this information can be considered guidance. Often the material is split between different but related documents; one containing the normative information and one or more the informative guidance.

A number of national and international standards are relevant to managing clinical risk in HIT. A useful Technical Report from the International Organization for Standardization summarises a selection of these [2]. A distinction is often made between those standards which apply to medical devices and those which apply to software which could impact care but falls short of the medical device inclusion criteria. It would be misleading for an organisation to claim compliance with a medical device standard for a non-medical device product. Nevertheless even for general HIT, much can be learnt about the principles of risk and quality management from the medical device standards (Table 4.1).

A. Stavert-Dobson, *Health Information Systems: Managing Clinical Risk*,
Health Informatics, DOI 10.1007/978-3-319-26612-1_4

Table 4.1 Lists out a selection of standards with potential relevance to HIT

Standard	Health IT specific	Medical device specific	Focus on risk or quality?
IEC 61508	No	No	Risk
ISB 0129/0160	Yes	No	Risk
ISO/TS 25238:2007	Yes	No	Risk
ISO 14971:2012	No	Yes	Risk
IEC 80001-1:2010	Yes	Yes	Risk
ISO 9000 family	No	No	Quality
ISO 13485:2003	No	Yes	Quality
ISO 62366:2007	No	Yes	Both
ISO 62304:2006	Yes	Yes	Quality

4.2 Examples of Risk and Quality Management Standards

4.2.1 IEC 61508

Prior to the 1980s the design of industrial control systems was largely the remit of the hardware engineer [3]. As microprocessors became more advanced and cheaper to manufacture, designers began switching to software to manage complex systems. But with this change came some interesting challenges. Engineers had been used to working with traditional electronics which could be proven using sound analytical techniques to behave in predictable ways. Establishing safety was largely an exercise in mathematics and simulation. When those same techniques were applied to software it soon became apparent that the methods had shortcomings and simply didn't translate; it proved difficult or impossible to prove with certainty that a software application was guaranteed to behave in a particular way. Indeed it was even questioned whether it was appropriate to rely on software for the management of safety critical operations at all.

The International Electrotechnical Commission (IEC) set up studies to investigate solutions to this growing problem. By the mid-1990s the makings of a standard was produced which introduced the idea of a risk-based approach to drive out specific safety requirements alongside general system requirements. By the year 2000 IEC 61508 [4] had been ratified and since then has been gradually adopted in a number of safety critical industries.

IEC 61508 was designed as a basic safety standard with a view to it being used as the basis for domain specific safety standards. It encourages those who adopt it to tailor the methodology to particular applications.

Some of the themes included in the standard are:

- The notion that safety is about far more than just reliability and that systems can operate in an unsafe manner even when they are working as designed.
- The need to apply risk management to all phases of the development lifecycle from vision to decommissioning.

- The concepts of hazards, controls, risk and risk acceptability.
- The defining of the term Safety Integrity Level (SIL), a number from 1 to 4 representing the risk reduction required for the safe operation of a particular functional component. SILs are essentially an analogue for the extent of rigour required during the development process.
- The notion that demonstrating that a system has undergone a safety assessment process is as important as actually achieving safety.
- The use of the ALARP triangle as the basis for risk acceptability (see Sect. 3.1).

IEC 61508 is not without its shortcomings particularly when one attempts to apply it to HIT. Firstly its origins in control system engineering requires a level of translation to take place to apply it to HIT. For a standard which is over 400 pages long this can require a significant investment of effort. Secondly the standard says little about human factors as a cause of risk. Operation of HIT is a careful balancing act between well-designed technology, trained users and a number of other extraneous factors. IEC 61508 focuses on the technological components but not the subtle complexities of human behaviour. Finally the standard stops short of suggesting the safety case as the primary vehicle for the articulation of the argument and evidence used to justify risk (see Sect. 11.3). Safety cases are becoming the mainstay tool of risk management in many industries and form the central discussion of this book.

4.2.2 *ISB 0129 and ISB 0160 (UK)*

The UK government in 2002 initiated a national project to deliver new, interconnected HIT services into primary and secondary healthcare organisations. This was to be implemented by a number of private sector Local Service Providers over a 10 year period. The UK National Health Service established the National Programme for IT (NPfIT) to manage the delivery of this £6 billion programme [5]. Concerned about the potential for these systems to introduce clinical risk, in 2004 NPfIT asked the UK's National Patient Safety Agency (NPSA) to conduct an appraisal of their approach to risk management. The NPSA concluded that NPfIT's approach was suboptimal and that they were failing to proactively manage risk in HIT systems effectively.

What the analysis had uncovered was an uncomfortable risk management gap – and one not confined to the UK. Products which were deemed to be medical devices were subjected to significant scrutiny and were generally required to comply with ISO 14971 [6] to meet risk management objectives. Those products which fell outside of the scope of being a medical device but which were still quite capable of adversely impacting clinical care rarely underwent any kind of risk assessment at all. Given that the majority of HIT systems in the UK National Programme for IT were non-medical devices and their implementation was quickly proliferating, action clearly needed to be taken.

In the absence of an appropriate sector specific standard, NPfIT initially referred suppliers to IEC 61508 [4] and required suppliers to develop a hazard register for their HIT products (see Sect. 11.2). A national risk management training course was set up for both suppliers and healthcare organisations to encourage good practice and widen awareness of the requirements.

By 2009 the UK National Programme for IT was being overseen by a body called Connecting for Health (CfH) and, working with the UK Information Standards Board, two new standards were drafted and approved:

- ISB 0129 (DSCN 14/2009) [7] – Clinical Risk Management: its Application in the Manufacture of Health IT Systems
- ISB 0160 (DSCN 18/2009) [8] – Clinical Risk Management: its Application in the Deployment and Use of Health IT Systems

These publications were based on standards ISO/TR 29322:2008(E) and ISO/TS 29321:2008(E) which had been drafted by CEN/TC251 to address the need for an International Standard for Clinical Risk Management of HIT systems. However the proposed Technical Specification and Technical Report failed to gain formal international ratification. Uptake of ISB 0129 and 0160 in the UK was slow partly due to their complexity and lack of a clear enforcement framework. In 2012 major revisions were undertaken with a focus on separating out requirements and guidance. The result was a set of four easy to use documents (two for manufacturers and two for healthcare organisations) with content similar to ISO 14971.

Some key features of the standards include:

- A formal handover of the manufacturer's deliverables at key points in the project to their customer providing a basis for the healthcare organisation to conduct their own risk management activities. NHS procurement policy in England requires healthcare organisations to incorporate compliance with the standard as a requirement in the commercial contract with the supplier.
- The ability to justify risk on an As Low As Reasonably Practicable basis (see Sect. 3.1).
- A structure for the identification of hazards, their mitigation and the production of CRM documentation.
- The need for both suppliers and healthcare organisations to appoint a Clinical Safety Officer who must be an experienced clinician with an appropriate professional registration. The Clinical Safety Officer is responsible for ensuring that a risk management process is formulated and adhered to. Importantly the requirement ensures that suitable domain expertise is included in the assessment and that deliverables are signed-off by an individual with professional standing. Of course the appointing of a Clinical Safety Officer is a challenge for smaller manufacturers who may not otherwise employ a registered clinician.

Importantly this pair of standards formally recognise for the first time the role of the healthcare organisation in formally managing clinical risk in HIT. Previous standards had focussed almost exclusively on the responsibilities of the manufacturer. With increasingly configurable systems being brought to market, it is not uncommon for the off-the-shelf HIT solution to be a toolkit ready for the healthcare organisation to configure into a working product. To ignore the risk which could be introduced at this phase of the implementation leaves a significant gap in the end-to-end safety argument. To expect the supplier to undertake this level of analysis is simply unrealistic and so ISB 0160 formally sets out the healthcare organisation's contribution to justifying the safety claims of the implemented product.

Although ISB0129 and ISB 0160 are primarily UK standards, interest is emerging internationally as recognition grows of the potential gap in managing the safety of HIT systems which are not medical devices. Nonetheless as the standards operate outside of any regulatory framework the challenge of enforcement, particularly against manufacturers, remains. Ultimately for any standard of this nature to succeed it will be for healthcare organisations to drive manufacturers to comply through the market forces of procurement rules and commercial contracts.

4.2.3 ISO/TS 25238:2007

ISO/TS 25238:2007 (Health informatics – Classification of safety risks from health software) [9] proposes a framework for assigning HIT systems into one of five risk categories. The standard is aimed at HIT products which are not medical devices but could lead to patient harm – as with many other standards in this area financial and other types of risk are excluded from scope.

The proposed categorisation is achieved by establishing categories for the severity and likelihood of harm occurring to a patient and then deriving a deriving a risk class through the use of a matrix. In this assessment the potential for harm that the product could present must be determined as if safety design features and controls are not present or are malfunctioning so the result is in effect a worst case initial risk. The standard proposes five risk classes A (Highest) to E (Lowest). The resulting measure of risk is intended to be used to drive the need for further control measures and the level of analytical rigour required. The standard stops short of suggesting risk acceptability levels.

ISO/TS 25238:2007 focuses on undertaking a risk assessment for a system as a whole rather than an exploration of individual hazards, causes and failure modes. This high level approach is useful at the early stages of a project where an 'order of magnitude' risk is used to inform the programme of work required. An analysis of this sort reveals a basis for the likely degree of effort and rigour required to assure the product without the need to consider the detailed design and technical architecture involved. Helpfully the standard sets out some examples of HIT systems with a proposed risk category and the associated rationale. However, in the absence of international standards for the development, deployment, configuration and use of HIT, this standard has seen little use. The long term future of ISO/TS 25238 and its relationship to more recent standards is currently unclear.

4.2.4 ISO 14971:2012

ISO 14971 "Application of risk management to medical devices" [6] is an international standard aimed at those products (including in vitro diagnostic devices) whose intended purpose falls within scope of medical device regulation. In practice the standard and supporting materials provide a useful framework for managing risk in HIT more generally, irrespective of the medical device status. The scope of the standard encompasses harm not just to patients but also to healthcare staff, property and the environment. ISO 14971 is itself referenced by a number of other related standards including ISO 13485 [10], ISO 62304 [11] and ISO 62366 [12].

ISO 14971 has been adopted by the European Committee for Standardization as EN ISO 14971:2012 [6]. Importantly it is one of the harmonised standards which manufacturers are obliged to comply with if they wish to certify their product as a CE Marked medical device (including in vitro diagnostic devices). Similarly in the US, the 21 code of federal regulations requires risk management to be undertaken as part of design validation (820.30 (g)). ISO 14971 is a Recognised Consensus Standard which the FDA expects manufacturers to comply with to meet this requirement.

The standard applies not only to physical medical devices but also to software products which meet the criteria. And to this ends an additional publication IEC/TR 80002-1:2009 [13] describes in detail how the requirements can be applied to medical device software. The standard requires that risk management activities are carried out throughout the lifecycle of the product, i.e. from vision through go-live and until decommissioning.

Key features of the standard include a need for the manufacturer to:

- Establish and document a risk management system
- Apply the risk management system to each medical device which is brought to market
- Systematically identify hazards and their causes
- Formulate, implement and validate the effectiveness of controls
- Evaluate the degree of residual risk and determine its acceptability
- Maintain the assessment within a risk management file
- Use competent personnel in the assessment of risk

Note that the international version of the standard and that used to achieve compliance with the EC Medical Device Directive vary slightly and in particular in the approach to risk acceptability criteria (see Sect. 3.3).

4.2.5 IEC 80001-1:2010

IEC 80001-1:2010 "Application of risk management for IT networks incorporating medical devices" is a voluntary standard being developed by the ISO/TC 215–IEC/SC 62A Joint Working Group. It has some similar requirements to ISO 14971 in that it calls for risk management processes, acceptability criteria and live environment monitoring.

However, the most fundamental difference is that it is aimed at healthcare organisations who are increasingly using general purpose networking solutions to connect highly regulated devices rather than using closely controlled proprietary networks supplied and supported by manufacturers. The approach can bring with it efficiencies, simplicity and the benefits of shared communication. However this needs to be balanced with the risks of connecting into an environment which is typically less well controlled, architecturally diverse and more dynamic in nature.

Depending on traditional hospital networks introduces new challenges and hazards for healthcare organisations. Factors include taking account of other traffic on the network, wifi constraints, reliability, in-life architectural changes, system upgrade management, security measures and remote access methods amongst others. Patient safety depends as much on the safety and secure exchange of medical information as it does on the safety performance of medical devices [14]. As such the standard aims to address the safety, effectiveness and security implications of using generic medical IT networks in this way.

Note that the standard is not intended to be used for assuring networks which provide the infrastructure for general HIT products which are not medical devices. Thus a network used only to support a simple Electronic Patient Record, email and internet access in a healthcare organisation would not typically be included within the scope of the standard.

IEC 80001-1 focuses on the lifetime management of the network including documentation requirements, change management, stakeholder engagement, etc. Healthcare organisations are required to define their medical IT networks and undertake a risk assessment before it is put into live service or when it is modified. The standard calls for personnel to be appointed to specific roles such as the Medical Network Risk IT Manager and for senior management to be held accountable for enforcing risk management. It requires stakeholders (medical device manufacturers, healthcare organisations, IT vendors, etc.) to collaborate under formal responsibility agreements which set out for the lifetime of the project what parties expect of each other in achieving compliance with the standard.

Specifically the standard avoids imposing any particular performance requirements or design characteristics on the network. Instead it encourages the justification of the chosen approach on the basis of risk. A number of associated technical reports have been published including guidance on distributed alarm systems and responsibility agreements. Although adoption of the standard has to date been limited, it is likely that awareness will grow in the next few years.

4.2.6 ISO 9000 and ISO 13485

Quality management is fundamentally different to risk management. Quality management is concerned with the production of a product or service which meets customer requirements and does so consistently. Quality management is implemented through the introduction of a quality management system – a set of agreed policies and business processes designed to achieve the agreed quality objectives. In contrast, risk management is about identifying and mitigating those factors which threaten the achievement of quality and other objectives. In clinical risk management we concern ourselves with those threats which may ultimately result in patient harm. Quality management pertains to the processes and procedures used to manufacture a product. Clinical risk management is generally concerned with the safety characteristics of a particular product.

Having said that, there are links between quality and risk management.

- Quality requires a certain degree of analytical rigour which, like risk management, may bring to light hazards which would have otherwise remained hidden
- Quality management calls for validation and verification, activities which may also provide evidence for controls in the hazard register (see Chap. 17)
- A culture of quality and continuous improvement may be associated with a similar ethos of assurance and risk management

The impact of quality management in an organisation can be far reaching in terms of its overall success. A well implemented system is likely amongst other things to:

- Raise customer satisfaction
- Drive increased revenue
- Provide efficiency gains
- Improve staff morale
- Promote a good reputation
- Reduce customer complaints
- Improve customer loyalty
- Avoid financial penalties
- Reduce the need for rework

ISO 9000 [15] has almost become synonymous with quality management in many industries. The standard was first published in 1987 and has since been adopted by over one million organisations in 180 different countries. ISO 9000 actually consists of a family of standards with ISO 9001 [16] containing the main requirements with which organisations must meet in order to claim compliance. Often organisations will choose to gain formal evidence of compliance by working with an auditor accredited with the ability to certify. Many purchasers require manufacturers to comply with the standard as a pre-requisite to doing business.

Most medical device regulators require that manufacturers implement an appropriate quality management system as part of their production process and in practice the requirements tend to be closely related to ISO 9001. The standard has been adapted to meet the requirements of several quality-critical industries including a version for the manufacture of medical devices. This adaptation is known as ISO 13485:2003 [10] and applies to the production of both software and physical devices. The standard is supported by technical guidance document ISO/TR 14969:2004 [17].

A number of regulators including those enforcing the European Medical Device Directive now prefer manufacturers to demonstrate conformity through compliance with ISO 13485. The standard has been harmonised with the EU MDD where it is given the title EN ISO 13485:2012. The compliance process usually involves engaging an accredited Notified Body who will, after undertaking a series of successful audits, issue a compliance certificate to the manufacturer. ISO 13485 contains a number of additional requirements above and beyond those of ISO 9001 which are specific to medical devices. In contrast some areas such as the monitoring of customer satisfaction are absent from ISO 13485.

ISO 13485 focuses on the management of quality throughout a product's lifecycle and sets out a framework for how organisations might tackle areas such as:

- Management's responsibilities, commitments to quality and process review
- Quality policy, processes and planning
- Gathering and review of requirements
- Document management, version control and record keeping
- The design and development processes, inputs and outputs
- Verification, validation, monitoring and test activities
- Feedback from customer organisations and the implementation of corrective actions

The standard stops short of any specificities for HIT, indeed it contains many requirements which may not be relevant (for example the need for contamination control and sterilisation). Therefore some degree of interpretation is always required and manufacturers would be well advised to discuss these with local regulatory advisors should their relevance be in any doubt.

4.2.7 ISO 62366:2007

ISO 62366:2007 "Medical devices – Application of usability engineering to medical devices" [12] is a design and development standard which is harmonised with medical device compliance requirements in many territories. It is closely aligned to risk management standard ISO 14971:2012, essentially ISO 62366 explores in greater depth those hazards which arise from a product's usability characteristics. Whilst intended as a standard to support the manufacture of medical devices it contains a useful framework that could be utilised by manufacturers of general HIT products where the user interface is associated with a number of important safety related hazards.

The standard defines usability as a "…measure of the effectiveness, efficiency, and satisfaction with which specified users achieve specified goals in particular environments, within the scope of the intended use of the Medical Device." The standard mainly focuses on safe operation rather than any potential improvements in efficiency or customer satisfaction.

ISO 62366 carefully uses the term 'use error' rather than 'user error' to highlight the fact that many errors are caused not by human carelessness but as a direct result of poor user interface design. It introduces the notion that such design flaws are important because they might only become apparent when the user's performance is compromised – in emergency situations or when fatigued.

The standard sets out a number of high level requirements outlining a usability engineering process. The process is aligned to a typical software development lifecycle and the activities are carried out in tandem with the product's design, build and test. ISO 63266 calls for early definition of the product's safety characteristics, these are established by the manufacturer based on the medical indication, intended patient population and operating principles [18]. From this information the hazards which relate to the user interface can be ascertained and analysed according to the process which is set out by ISO 14971. There is also a need to identify the product's primary operating functions, i.e. those user interactions which occur commonly and repeatedly and are therefore likely to pose a greater risk should they be used incorrectly.

This information, including the identified hazards are then embodied within a usability specification – the user interface requirements on which the product is designed and built. In formulating the specification the standard requires the manufacturer to consider both normal use of the product (i.e. when it is being operated correctly within its intended purpose) and during conditions of reasonably foreseeable misuse. Misuse refers to those actions taken by a user or organisation which are not intended by the manufacturer but might to some extent be expected in the target operating environment.

During and after the build process ISO 62366 then requires the user interface to be tested and verified against the usability specification. Where the product fails to meet the usability requirements this may act as the seed for a change in the design. This forms the basis of an iterative usability cycle where the product or a prototype is subject to constant verification by individuals who are representative of the intended user-base.

The output of the usability activities form the usability engineering file and the standard suggests that this forms part of the wider risk management file. The material is typically subject to audit by the Notified Body certifying the medical device.

At the time of writing ISO 62366:2007 is due to be replaced by a new version of the standard, ISO 62366-1:2015 and its accompanying ISO 62366-2.

4.2.8 IEC 62304:2006

IEC 62304:2006 (Medical device software – Software life cycle processes) [11] is harmonised with a number of regulatory frameworks and sets of the requirements for a software development lifecycle. The scope includes software which is integrated into a medical device or which stands alone. The standard sets out a series of requirements culminating in a rigorous methodology for software development from conception to live operation. Which requirements need to be complied with depends on an A, B, C categorisation reflective of the risk the product poses to the patient.

The framework includes the need to formally gather software requirements, develop architectural designs, establish risk control measures, verify software through testing and establish how the software will be maintained and changes managed. The standard also contains guidance on the complex area of integrating Software Of Unknown Provenance (SOUP) which may not have been developed with the same degree of rigour as the primary product.

At the time of writing work has started on a second edition of IEC 62304.

4.3 Summary

- There are a number of standards relating to risk and quality management which have implications for managing clinical risk in HIT.
- The UK standards ISB 0129 and ISB 0160 are the only known examples of standards that are actively used to specifically manage clinical risk within HIT which is not a Medical Device.
- Many of the international standards relate specifically to medical devices and these are not always suited to HIT. This gap is the subject of on-going work within International Standards Committees. Nevertheless, a great deal can be learnt from the principles set out in Medical Device standards.

References

1. British Standards Institution. British Standards Institution website. [Online]. 2015. [cited Aug 2015]. Available from: http://www.bsigroup.com/en-GB/standards/Information-about-standards/what-is-a-standard/.
2. International Organization for Standardization. ISO/TC 215 N1143 – ISO/DTR 17791 – Health informatics—guidance on standards for enabling safety in health software. 2013.
3. MTL Instruments Group plc. An introduction to functional safety and IEC 61508. Application Note. AN9025-3. 2002.
4. International Electrotechnical Commission. International Standard IEC 61508: functional safety of electrical/electronic/programmable electronic safety related systems. Geneva: 2000.
5. Campion-Awwad O, Hayton A, Smith L, Vuaran M. The National Programme for IT in the NHS. A case history. 2014.
6. International Organization for Standardization. EN ISO 14971:2012. Medical devices. Application of risk management to medical devices. 2012.
7. Health and Social Care Information Centre. ISB0129. Clinical risk management: its application in the manufacture of health IT systems version 2. UK Information Standard Board for Health and Social Care; 2013.
8. Health and Social Care Information Centre. ISB0160. Clinical risk management: its application in the deployment and use of health IT. UK Information Standard Board for Health and Social Care; 2013.
9. International Organization for Standardization. ISO/TS 25238:2007 (Health informatics – classification of safety risks from health software). Technical specification. 2007.
10. International Organization for Standardization. ISO 13485:2003. Medical devices – quality management systems – requirements for regulatory purposes. 2003.

11. International Organization for Standardization. IEC 62304:2006. Medical device software – software life cycle processes. 2006.
12. International Organization for Standardization. IEC 62366:2007. Medical devices – application of usability engineering to medical devices. 2007.
13. International Electrotechnical Commission. IEC/TR 80002-1:2009. Medical device software, Part 1, guidance on the application of ISO 14971 to medical device software. 2009.
14. ECRI Institute. 10 Questions about IEC 80001-1: what you need to know about the upcoming standard and networked medical devices. [Online]. 2010 [cited 17 Mar 2015]. Available from: https://www.ecri.org/components/HDJournal/Pages/hd201005p146guid_iec80001.aspx?tab=1.
15. International Organization for Standardization. ISO 9000 – quality management. 2000.
16. International Organization for Standardization. ISO 9001:2008 quality management systems. 2008.
17. International Organization for Standardization. Medical devices – quality management systems – guidance on the application of ISO 13485:2003. 2003.
18. North B. The growing role of human factors and usability engineering for medical devices. What's required in the new regulatory landscape.

Part II
The Causes of Risk in Health IT

Chapter 5
Safety and the Operating Environment

5.1 Health IT and the Socio-technical Environment

James Reason famously described the path from hazard to harm as being likened to a journey through Swiss cheese [1]. Each slice of cheese represents a step in a process in which there is a weakness or hole – an opportunity for failure. When slices are stacked together the holes generally fail to align. Even if one penetrates an individual hole the serial placement of subtly different barriers will be sufficient to hinder complete passage through the cheese. Each cheese slice acts as a defence, the more slices there are, the less likely it is that the holes will align.

HIT systems in live service rarely cause harm in isolation; instead they form just one component (a cheese slice or hole) in a complex ecosystem in which human operators co-exist and interoperate with technology. This situation is often referred to as a socio-technical environment. An understanding of this environment is important as scenarios which lead to harm also evolve from these relationships. Incidents are nearly always multifaceted and involve the interplay of historical design decisions, operating procedures, system architecture, training strategies and so on.

It is also naïve to conclude that any non-technical influential factor in contributing to risk lives exclusively in the domain of 'use error'. Whilst it is the case that all humans are fallible and perhaps even prone to error, to take this to the extent of attributing blame in the risk management process is largely unhelpful. In the main humans make mistakes not because they are reckless or malicious but because they operate in an environment which allows or even facilitates those errors being made. This view should therefore make us think differently about how we mitigate risk. Even if we succeed in temporarily correcting human behaviour, whilst ever adverse conditions prevail we continually require of those around us to operate on a knife edge. Therefore mitigating risk through analysis of the technology and human error alone is short-sighted.

Safety emerges from a careful evaluation of all the ecosystem's components and complex interactions. Only when these are understood can we begin to tackle the

A. Stavert-Dobson, *Health Information Systems: Managing Clinical Risk*,
Health Informatics, DOI 10.1007/978-3-319-26612-1_5

underlying causes of risk. Research and experience has shown that outside of healthcare great progress has been made by addressing the conditions that provoke error rather than simply telling people to be careful [1]. Human error may sometimes be the factor that immediately precipitates a serious failure, but there are usually deeper, systemic factors at work which if addressed would have prevented the error or acted as a safety-net to mitigate its consequences [2].

In 2010 Sittig and Singh suggested a socio-technical model proposing a set of human and technological elements which collectively contribute to the design, deployment and operation of a HIT system [3]. Similarly, the Systems Engineering Initiative for Patient Safety (SEIPS) model has been developed as a framework for better understanding the structures, processes and outcomes in healthcare [4]. A socio-technical approach to HIT has a number of proponents and the alternative, a 'techno-centric' model, has been cited as being responsible for the limited success of some major HIT programmes [5].

An adapted set of components from of Sittig and Singh's ecosystem is set out below:

- Hardware and software – constitutes the physical infrastructure, computer code, user interfaces, operating system and other software dependencies which make up the technical system. These components will typically be assembled according to a strategic architecture drawn up by the manufacturer or healthcare organisation and designed to balance cost, performance, functionality and resilience.
- People – represents the full set of stakeholders involved in designing, implementing and operating the system. Note that this will include the project sponsors, financiers, policy makers, programmers, testers, trainers, project managers, etc. Also included is the knowledge and experience that those people bring to the environment which they call upon to make clinical decisions and operate the system correctly.
- Workflow – embodies the clinical business processes in which users operate in order to deliver care as well as the procedures to enter, process and retrieve data from the system. The workflows may be formally documented and agreed or exist culturally as a result of training and experience. Established local workflows of course may or may not have been taken into consideration in the design of a particular HIT system.
- Organisation – encompasses the decisions, policies and rules which an organisation sets in delivering care and implementing HIT systems. These include the funds and other resources which may be available to support the HIT system. The organisation will also typically set the strategic goals of the system, the long term ICT approach and areas of priority. A key characteristic of the organisation in terms of risk management is the extent to which it manages to forge a culture of safety in the delivery of care and implementation of HIT systems.
- Content – Many systems contain not just the data which has been captured in the course of delivering care but also non patient-specific content which powers application functionality. For example, where a system offers drug-drug interaction or dose-range checking the system needs to maintain the relevant rules and

logic to contextualise this advice. Similarly, HIT systems often use standard vocabularies such as SNOMED CT, national catalogues of orderable items and other reference data.

- External influences – Healthcare organisations and system suppliers are influenced by the wider political, cultural, economic and regulatory environment in which they operate. Health policies and objectives change over time and vary geographically. These changes influence local priorities and decision making which impacts the kinds of system which are required, the necessary functionality and the resources available to support procurement, implementation and maintenance.

To demonstrate how the different components of the socio-technical model can interact a fictitious but not atypical user story is set out below:

> Dr Green working in a high-volume acute hospital prescribes an in-patient an anticoagulant using a recently implemented ePrescribing solution. The prescription spawns a task to administer the medication at 11:00 am and this is successfully given by the nurse. Unfortunately Dr Green did not realise at the point of prescribing that the patient in question was already receiving a different anticoagulant which should have been discontinued. The patient suffered a gastrointestinal haemorrhage as a result of the duplicate therapy.

One could argue that this is a simple case of human error or poor judgement – Dr Green should have checked what medications the patient was taking at the time of prescribing and taken this into account. By this logic the fault lies purely within the 'People' component of the socio-technical model. In fact we can go so far as to attributing direct blame in this case – a single individual with an accountable duty of care acted unprofessionally and contrary to best practice. The HIT system is safe, it merely executed an operation instigated by someone who frankly should have known better.

Investigating this fictitious system further reveals that it is capable of displaying all the medications that a patient is taking to make the appropriate check. However in this case, the drugs happen to be split across several screens. One screen shows regular medications, another those which are administered on request (PRN) and yet another for infusions. Several clicks are required to visit each screen and there are no indicators as to whether the screens contain anything prior to them being accessed (a contribution from the 'Hardware and software' component of Sitting and Singh's model). In this case Dr Green omitted viewing the 'Infusions' page where the existing anticoagulant was listed. So whilst Dr Green is technically able to see the information he needs, doing so is not necessarily made easy ('Workflow' component).

The reason for medications being split over several screens turns out to have its origins in the system requirements. These specified that the system must "Clearly identify which medications are administered regularly, which on request and which are infusions." The specification authors failed to be explicit or state particular safety requirements for the user interface ('People' component). The system designer took that to mean that several screens should be created. Unfortunately at the time of designing the system senior managers at the hospital were unable to free clinical time for experts to review and contribute to the subject matter ('Organisation' and 'Workflow' components). As a result of a reduction in government funding, several members of clinical and technical staff had recently left the organisation and a decision had been made not to replace them ('External' component).

The system happened to have functionality to detect when potentially conflicting medications were prescribed but in this case no warning had been offered. Earlier that month a decision had been taken by system managers ('People' component) to disable the functionality as users were being overburdened with irrelevant warnings and developing alert blindness ('Hardware and software' component). The rules governing drug interaction checking were not specific enough and were generating false positives ('Content' component). Unfortunately the fact that the functionality had been disabled had not been communicated to users ('People' component) and the software did not display an indicator as to whether the checking was active or not ('Hardware and software' component). Users would have reasonably expected the system to provide this validation as it had done in the past ('Workflow' component).

With just a small amount of context and investigation it is possible to develop a rich picture of the many contributing factors to this incident. To pretend that one can ascertain a single root cause is at best naïve. If we insist on putting this down to simple human error how would one look to mitigate that in the future? Disciplinary action against Dr Green maybe, perhaps with a requirement to undertake more system training? Such a blunt tool fails to address the underlying factors which lead to the toxic environment in which personnel are operating in this story. Only by addressing the bigger picture would it realistically be possible to proactively manage risk in the long term.

So how might this scenario have turned out differently if a risk management system had been embedded into the development lifecycle? Typically design reviews, hazard workshops and static testing would have identified the ambiguous user interface requirement and corrected the design long before build took place. By involving a multidisciplinary team in the early stages a hazard would have been detected and efforts to reduce the risk to as low as reasonably practicable would have generated a plan to address the problems. System testing should also have identified both the user interface and poorly performing drug interaction checking. This would have spawned appropriate corrective actions and the risk of not addressing them would have been established early in order to ensure optimal prioritisation.

It might have been appropriate to eliminate the interaction issue by switching the functionality off before go-live rather than after. This would have set the expectations of users from the outset and avoided the difficulty of modifying otherwise entrenched workflows. The decision to switch off drug interaction checking during live service should have been managed under formal change control processes. This would have identified that doing so would either create a new hazard or highlight the removal of a control to an existing hazard. Further mitigation would have been required and this might have included the need to formally communicate the change (via a number of different media) to users.

The point is that this scenario could have been foreseen early in the project if only the right analysis had taken place and potential risks taken into account. Dr Green was not to blame he just happened to be the first unfortunate victim.

Essentially the organisation was asking him to mitigate high level institutional failings and to do this without warning during a busy ward-round after a night on call. This is, and proved to be, an unrealistic human factor mitigation. What we learn is that to describe a system in isolation as being safe or unsafe is, at best, to oversimplify. Such a binary and restrictive measure fails to address the notion that safety is a property which emerges from the interaction of the ecosystem's diverse components.

One final point is that this scenario is an example of risk existing in a system which is functioning under normal rather than fault conditions. No amount of fault-finding purely at the software level would have identified these hazards.

5.2 Building a Safety Culture

A key characteristic of an SMS is that it is integrated into day-to-day business in such a way that most or all employees within an organisation operate within it. In this way safety becomes a collective responsibility where anyone can detect hazards, single points of failure and other threats. But developing processes for employees to follow is only part of the picture. Encouraging an awareness of safety and positively rewarding the effective handling of risk is an important responsibility for top management. Mark Rosenker, Chairman of the National Transportation Safety Board claims that an SMS "…necessitates a cultural change in an organization so that the safety of operations is the objective behind every action and decision by both those who oversee procedures and those who carry them out" [6].

Bayuk [7] points out that "Employees must fully trust that they will have management support for decisions made in the interest of safety, while also recognising that intentional breaches of safety will not be tolerated. The result is a non-punitive environment that encourages the identification, reporting, and correction of safety issues." Developing such a non-punitive culture is a challenge which both healthcare organisations and manufacturers need to embrace.

The importance of safety culture and its ability to provide an environment where we learn from our mistakes is beginning to percolate through healthcare and system manufacturers. In the UK Department of Health's publication An Organisation With a Memory [2] the authors point out that organisational culture is central to every stage of the learning process – from ensuring that incidents are identified and reported through to embedding the necessary changes deeply into practice. It goes on to acknowledge that blame cultures on the other hand can encourage people to cover up errors for fear of retribution and act against the identification of the true causes of failure.

Many would say that in the last 10 years a cultural revolution has occurred in healthcare delivery. Organisations in many countries are now more aware than ever of the potential for healthcare to cause harm. In response, most healthcare organisations have put in place measures to identify, report and manage adverse incidents and near-misses along with proactive risk control processes. The challenge for managers of these organisations is to extend this cultural awareness into the implementation of

HIT by harnessing existing processes but with recognition of the novel complexities which technology brings.

For many manufacturers achieving a solid safety culture from a standing start is a more difficult challenge. Often HIT vendors exist as part of larger organisations that traditionally operate outside of healthcare in lower risk domains. A number of drivers usually form the foundations on which to build a safety culture; an awareness of the benefits of CRM, added product value, threat of legal challenges, the need to comply with mandatory standards or regulations and frankly a drive to do the right thing.

A safety culture can be driven through the combination of:

- A clear message from top management (usually in the form of a CRM Policy) – see Sect. 8.4
- The introduction of robust processes via an SMS – see Sect. 8.1
- The regular delivery of CRM training to employees to raise awareness of the policy and processes – see Sect. 8.10
- Ownership of risk by front-line staff
- The reporting and analysis of adverse incidents and system defects to drive continual improvement
- Visible leadership and oversight from the highest levels of an organisation

5.3 Human Failure

The clinical environment can be chaotic at times. Clinicians are required to mentally integrate complex information from many different sources whilst being subject to frequent interruptions, distractions and time pressures. HIT has the potential to bring some order to that chaos but those same factors and the need to multitask can also threaten safe operation of the system. The human-technology interface is critical to assisting users in making the right choices and an exploration of the manner in which people fail is worthy of some thought. In particular when it comes to formulating the hazards which a HIT system might present, it is useful to have a framework to help us appreciate the modalities by which users may act in error.

Human failure can essentially be split into two categories; circumstances in which our actions are unintentional and those which are deliberate. Inadvertent actions constitute human error whilst deliberate activities fall into the remit of violation.

5.3.1 Human Error

Cognitive psychologists describe different types of human performance [8]:

- Skill based performance – those actions which we undertake with minimal cognitive effort because they are highly practiced. To all intents and purposes these are activities which are automatic or subconscious.

- Rule based performance – behaviours which require more mental effort and planning because we are operating in less familiar territory or where a skill-based approach isn't going to work.
- Knowledge based performance – activities where we have little pre-existing skills or rules to call upon and we need to assimilate the facts from first principles using inference and judgement.

Human error is defined by James Reason as "the failure of planned actions to achieve their desired ends – without the intervention of some unforeseeable event" [1]. Depending on the mode in which we are performing, different types of errors might occur. These are often referred to as slips, lapses and mistakes.

Slips and Lapses

Slips and lapses describe those errors where an individual selects the correct action but executes it in a manner which fails to achieve the intended objectives. For example, we might habitually add sugar to a colleague's cup of coffee when the recipient doesn't actually take it. Slips and lapses tend to occur when we are undertaking skilled, automatic activities in familiar situations. In these circumstances our attention is called upon only to monitor our actions and make corrections as required. Slips and lapses tend to occur at times when we are distracted. Often slips are a result of us re-using learned behaviour in a slightly different but inappropriate context.

Psychologists often refer to different forms of slips. A capture slip occurs when we begin executing a particular activity only to be diverted into doing one which is more familiar. For example, getting into the car and automatically driving to work rather than destination one intends. A description slip happens when we confuse two similar tasks for example, putting a bottle of milk away in the cupboard rather than the fridge.

Examples of slips in the context of HIT might include:

- Accidentally clicking a 'save' button when we meant to press 'cancel'
- Selecting the wrong drug from a list of similarly named drugs
- Referring a patient to a commonly used clinical service rather than the one they actually require

Lapses are similar to slips but pertain to situations where we omit an action or forget what our original intention was. Lapses are therefore related to memory, examples might include:

- Beginning navigation to a particular screen but becoming waylaid and ending up reading information in a different screen
- Intending to order several investigations but only ordering one
- Typing information into the previous patient's record because the user has forgotten to bring the next one into context

Slips and lapses often occur when the task being undertaken is overly complex or long-winded. Similarly we may be susceptible when we are dealing with steps which fail to flow naturally or when we need to remember to do something rather than it being a logical part of the workflow.

Interestingly even the most experienced and highly trained personnel are susceptible to slips and lapses. Individuals know full well what they are trying to achieve and how to achieve it but distractions in the environment or personal stressors win over and compromise their performance. For these reasons, additional training rarely reduces the incidence of slips and lapses. Instead a combination of good system design, effective processes and the use of tools such as checklists and confirmation screens guide us into taking the correct actions.

Mistakes

Mistakes are errors of systematic decision making and occur when we initiate the wrong action but believing it to be correct – essentially this is the failure of planning [9]. We make mistakes when we undertake tasks which are less familiar and require more cognitive effort than for skill-based activities. Often the behaviour is a result of us being overburdened with matters competing for our attention.

Mistakes can be either rule based or knowledge based. When we have some but limited experience of a situation we may call upon mental rules to decide on the best course of action. In these circumstances there is the possibility that we may select the wrong rule for the given set of circumstances or select the correct rule but wrongly apply it. When we use knowledge to govern our behaviour we are reliant on having sufficient information to make a judgement and also to be able to assimilate and interpret the information correctly to reach the best unbiased conclusion. At times this process may fail resulting in a mistake. Examples of mistakes in the operation of HIT might include:

- Scheduling an appointment on a day when the clinician is unavailable
- Selecting a care pathway that is inappropriate for the patient
- Recording a firm diagnosis for the patient when in fact the situation is still unclear
- Overly relying on decision support advice provided by the system which was actually inappropriate given the circumstances

5.3.2 *Violation*

Violation is distinct from error in that motivation plays a greater part than knowledge. Violation is a deliberate act, an action which the perpetrator knows he/she should not undertake but does so regardless – essentially risk-taking behaviour. The activity in some way contravenes existing wisdom which might be in the form of policy, training, professional guidance or best practice. Although the individual may intend to commit the violation the consequences are often unexpected.

Violations can be categorised further [10]:

- Unintentional – for example as a result of rules which are impossible to follow in practice.
- Routine – where individuals habitually contravene best practice, for example taking shortcuts or clicking through warning screens without reading the content. Violations in this category are often commonplace, institutional and condoned by colleagues.
- Situational – where the circumstances mean that compliance with good practice is difficult, inconvenient or time consuming.
- Optimising – where individuals believe that they will perform better by cutting corners and not complying with good practice.
- Exceptional – where people are of the opinion that the rule simply doesn't apply to them or their current circumstances

Violations can be particularly hazardous when they are combined with error. The violator knows that he is acting incorrectly and is therefore more likely to hide his/her behaviour.

The extent to which individuals violate can to some extent be linked to safety and governance culture within an organisation. People are more likely to violate the rules if they function in an environment where rules are frequently bent, where people feel the need to take shortcuts or exhibit 'macho' behaviour. Simply forcing individuals to comply by issuing threats is unlikely to be effective in the long term. Careful analysis can reveal that violations occur because individuals feel that they have no alternative means of achieving their objectives, that compliance requires an unworkable business process or that they simply don't see the point. In some cases a violation may prove to be the seed of new requirements or modified ways of working. Usually the best way of managing a violation is to remove the situation which causes the individual to violate. At the very least, human are more likely to comply when they understand the rationale for rules being in place.

5.4 Human Error and User Interface Design

Usability is the characteristic of the user interface that establishes effectiveness, efficiency ease of user learning and user satisfaction [11]. Usability impacts not only safety but productivity, user fatigue, user take-up, customer acceptance and satisfaction. Poor usability of HIT is widely believed to have a substantial negative effect on clinical efficiency and data quality [12].

Many industries including nuclear engineering, aviation and the military have come to realise that human-system interaction can have a major impact on safety. In these domains regulatory processes have been put in place to mandate usability evaluation. Both the US FDA rules and Europe's Medical Device Directive require manufacturers to undertake usability studies (see Sect. 4.2.7). Even for products outside of medical device regulation, the contribution of usability issues to patient safety is significant enough to warrant specific risk assessment. In a study of over 3,000 US group practices, Gans found that usability was a significant barrier to Health IT uptake [13]. Some authorities have suggested that poor usability is one of the major factors, possibly the most important factor, hindering widespread adoption of Electronic Medical Records [14].

5.4.1 Design Methodology

Designing an optimal user interface is a careful balance between many different factors. A sound user interface is not about the quality of the graphics or the colour pallet chosen, rather it is about its effectiveness, the ease and efficiency with which users can achieve their objectives safely and consistently.

The five goals of user interface design are often cited as:

- Learnability: How easy is it for users to feel comfortable using the system the first few times it is operated
- Efficiency: The ability of users to achieve their objectives and to do so in a timely manner
- Memorability: The extent to which users are able to continue being proficient in using the system when they return to it after some time
- Errors: How frequently users make errors in operating the system
- Satisfaction: The extent to which the system is aesthetically pleasing and provides a sense of fulfillment in its operation

Of course not all users prioritise these characteristics equally. An expert user is more likely to focus on efficiency whilst a beginner will require learnability and memorability. Errors in safety-related HIT systems frequently translate into the propensity to trigger hazards. On this basis, all HIT system designers need to prioritise managing error occurrence as a design objective.

Products with suboptimal user interfaces can be difficult and expensive to fix once they are in live operation. Not only do changes need to be carefully implemented but training materials and user documentation are also likely to need a refresh. It is far more efficient to get the user interface right from the start. Similarly manufacturers need to have an appreciation that poor interface design cannot be realistically be remedied by additional training. As discussed in Sect. 2.6 most slips and lapses simply cannot be addressed through the provision of advice or guidelines.

Sound user interface design requires engagement and close cooperation of representative users. In practice it is insufficient to simply hold a focus group to ask users what their requirements are. A better approach is to formally implement a user-centred design process. Here a small number of representative users are convened and comments are solicited on candidate designs. As the project progresses designers need to offer prototypes or test environment for the subjects to try out the product.

Users should be encouraged to 'play' with the software whilst being carefully observed; what does the user find easy to do, what is more challenging, how do they feel as beginners or experts? Users should be encouraged to go off-piste in their evaluation and interact with the system in unpredictable ways which engineers may not have considered. Evaluators should observe how easily users achieve their objectives, whether they deviate into irrelevant screens or if they need to opt for a suboptimal solution to their problem.

An iterative, user-centred approach has advantages in many areas of product development but perhaps none more so than in user interface design. Manufacturers should actively plan to work and re-work a user interface many times in the course of a project. Of course the earlier this process begins, the more opportunities there are to tweak the design without compromising project timescales. In the context of a time-bound software development project it may not be realistic to evaluate every screen and function in a system so it makes sense to prioritise those which are used frequently and/or are critical for safety. Even after go-live it is invaluable to solicit feedback from users once they have had the time to use the system in anger and address areas of concern.

In some cases engineers may be able to draw on previous experience or established best practice in formulating safe designs. Important lessons may have been learned in previous versions of a product, from alternative products or as a result of analysing incident reports. Issues of intellectual property and patent law aside, successful user interface components are frequently replicated in the technology industry not only leveraging experience but introducing consistency in the process. Some attempts have been made to standardise user interface components in HIT to improve the user experience for those clinicians who operate across several organisations. Most notable is perhaps the UK's Common User Interface Project undertaken in partnership with Microsoft [15]. The project had some success in standardising the way that critical data such as the patient identifying banner, demographic data and allergies are presented. Similarly the NHS's 'Guidelines for safe on-screen display of medication information' has made inroads into bringing consistency to the way that drugs are presented and described [16].

5.4.2 Minimising Errors by Design

In designing a safe user interface, the engineer should take into account what we know about human error and cognition to forge the underlying design principles for a system. Applications which disrupt rather than encourage safe workflow, require users to keep information in their head from one screen to the next or fail to provide the right tools for the task in hand are likely to increase risk. Those which are sympathetic to the way users think and take into account what we know about human error will not only control hazards effectively but are more likely to be accepted by users. Some simple design rules (such as those proposed by Shneiderman et al. [17]) implemented early in a project can greatly assist in mitigating user interface hazards.

Avoiding Slips

In Sect. 5.3.1 we discussed the concepts of slips and lapses and there are a number of design principles which can be employed to reduce the incidence of these. Slips for example can be addressed by avoiding common but not identical action sequences. Suppose viewing or cancelling an appointment requires an almost an identical series of mouse clicks. Should one of these activities be much more common than the other it is likely that from time to time an error will be made when actioning the infrequent task. Similarly buttons with unrecoverable consequences should be separated from more benign, frequently used functions. For example, it would be wise to design in some physical distance between buttons entitled "Record patient death" and "Update patient address".

It should be clear by observing a button or icon what action it performs prior to it being activated. Ergonomic engineers use the term 'affordance' to refer to a quality of an object which allows an individual to perform an action. For example, in the physical world one would expect a knob to work by twisting it, a cord with a bobble on the end by pulling it. In these circumstances we don't require training or extensive experience with an object in order to use it correctly, the mode of operation is obvious by nature. Well-designed user interfaces will take account of affordance and offer the best control to the user to meet the task in hand. For example, if a user is required to select a single option from a series of options then a radio button might be a suitable choice. Requiring the user to type Yes or No into a text box would be a much less elegant solution, prone to human error and possibly a source of risk. It is also valuable to provide feedback to the user when an action is detected and completed to prevent them re-initiating. For example a colour/caption change on a button after it is clicked or the use of a continually moving progress bar to monitor slowly executing processes.

A mode error is a particular kind of slip which has importance in technological systems. These occur when we believe a system to be in a particular state when in fact it is not. For example, a password may be repeatedly rejected because the user has accidentally pressed the caps lock key. The 'caps on' mode renders the system into a state which was unexpected, unforeseen or forgotten. The use of modes in technology is often helpful as it reduces the command set and can produce a more compact user interface. However when relied upon in a safety critical business process, modes can present a significant hazard.

A common example of a mode error in HIT systems is the use of filters to rationalise lists of data. Lists are commonly employed in user interfaces to create a master-detail workflow; the list provides an overview of the available data which is linked to a second screen showing the complete dataset. Where lists have the potential to be long, designers sometimes provide a filter control. However clinicians use lists not just to ascertain what information is available but to establish what is missing. Suppose a system lists the immunisations which have been administered to a particular child. A clinician might scan the list to determine any which have been missed. Of course should a filter be inadvertently applied, wrong assumptions could easily be made. Designers who foresee this hazard will ensure that the presence and nature of a filter (i.e. the current mode) is clearly indicated and that the default setting on entering the screen is for the filter to be cleared.

Avoiding Lapses

Psychologists use the term ‘cognitive load’ to refer to the brainpower and reasoning needed to undertake a given task. The term can be defined as “the extent to which cognitive resources, especially working memory, are utilized during learning, problem solving, thinking, and reasoning” [18]. The higher the cognitive load the more taxing it is to do something. The clinical environment is full of distractions and situational awareness requires a careful balance between the supply and demand of attentional resources. Stress, time pressure, and lack of sleep are but a few of many factors which can decrease the resource pool [19]. Research has shown that HIT is generally perceived as reducing overall cognitive effort amongst clinicians [20]. Nevertheless poorly designed user interfaces can add to the cognitive load in many cases unnecessarily.

Delving more deeply, ‘intrinsic cognitive load’ represents the effort needed to absorb the information we are looking at and make clinical decisions, on the other hand ‘extrinsic cognitive load’ is the additional thinking power needed to understand the user interface and operate the system itself. It is extrinsic load that designers endeavour to minimise by providing a user interface which is intuitive.

Interfaces add to cognitive effort when they:

- Require users need to remember information from one screen to the next
- Force complex decision making
- Leave the user unclear as to what the next step in a process is
- Require navigation from one screen to another to assemble the necessary information
- Necessitate hunting for information in screens which are information dense

A particularly common type of lapse is known as the Post Completion Error (PCE). These occur when the subject feels that they have concluded a particular workflow when in fact there are one or more outstanding stages to complete. For example, in the physical world one might fail to replace the petrol cap after refuelling or accidentally leave an original document on the scanner of the photocopier. In HIT the user might read a laboratory result but fail to formally acknowledge the task as complete or document some clinical notes without forwarding them on to an intended recipient. The error often occurs when an individual is forced to specifically remember the final activity in the absence of any visual clues to provide a prompt.

To minimise lapse errors designers need to focus on workflow. An interface which mirrors the logical sequence of its users’ thoughts will make it more difficult for them to accidentally deviate from their original goals. Sequences should have a well-defined beginning, middle and end with clear indicators of completion. Keeping workflow short and the relevant information accessible at the right time will guide users into taking the right path by default. For some critical areas it might be appropriate to design the system to enforce a particular workflow where the system insists that users complete tasks in a particular order. This technique better facilitates completeness at the expense of constraining the user's control.

Avoiding Mistakes

Mistakes are planning errors, their causes are multifactorial and include training, experience and application knowledge. A well-designed user interface will guide users into making the right choices most of the time, in other words the right thing to do should be the easiest. Where users make incorrect choices the user interface should be designed to aid detection of this and, where possible, offer the opportunity to reverse an action. In general, 'undo' functionality is far preferable to endless requests to confirm actions.

Users will quickly construct mental rules as they operate the system for the first time. The assimilated rules empower them to go on to proficiently operate unfamiliar parts of the application using the conventions learned in more accustomed areas. To achieve this designers need to strive for consistency in user interface design. For example, if some screens contain buttons labelled 'Save' and 'Close' whilst others have 'Okay' and 'Cancel' for the same functions, this is likely to prove confusing and hinder rule development. In fact one of the least technically challenging enhancements to a user interface is to simply ensure that material is labelled appropriately, that abbreviations and ambiguities are avoided and button captions reflect their true function.

Error messages are seemingly benign in the grand scheme of designing the user interface but their effective use can have an impact on minimising slips and lapses. Careful thought should go into crafting how messages are worded. For example, "Invalid data, unable to perform action" is not terribly helpful in solving the user's problem. Messages are more effective when they are specific, use natural language and suggest how the problem might be solved. For example, "You entered a date of 12/09/2013 for the next appointment but this is in the past. Check that you are using the correct date format dd/mm/yyyy".

The effective and consistent use of colour in the user interface can greatly aid the identification of relevant information. However special attention should be given as to how screens might appear to those users who are colour blind. It is estimated that around 8 % of males have some degree of colour perception impairment. Designers need to take this into account when displaying safety critical information especially when it comes to indicating status. For example, should a system use colour coding alone to signify whether or not a numerical result is outside the normal range, this could prove difficult to interpret for colour blind users. As a general rule, if the status of information is signified by a colour change then this should be supplemented by some other indicator (e.g. bold text, an icon or a character such as '*').

Finally thought should be given as to how priority information should be handled in the user interface. The density of information can impact the ability to rationalise and prioritise what is important and designers should avoid cluttering screens with irrelevant information. Some data items will be critical to the delivery of safe care but drawing attention to key information without it being intrusive to the workflow can be a challenge. A popup box declaring "User allergic to Penicillin" on accessing a patient record is likely to catch the eye but the continued distraction from the task in hand may lead the user to automatically click 'Okay' without it ever being truly noticed. Placing this information next to where a drug is selected for prescribing is more likely to offer a better means of detection.

Difficulties are also encountered when one deals with information which is safety critical to one clinician but may be almost irrelevant to another. For example, "Patient difficult to intubate" is a life-saving nugget of data for an anaesthetist but may be no more than a waste of screen real-estate for that patient's dermatologist. User interfaces which allow a degree of role-based configuration can sometimes be the optimal solution in these cases.

5.4.3 Mental Models and Predictive Behaviour

Psychologists use the term 'mental model' to describe the conceptual picture of the system we carry in our minds. Our mental model has major influences on the choices we make when we operate a system. We apply the model to make predictions about how the system will behave in unfamiliar circumstances. A well designed system will induce in the user's mind an accurate picture purely through operational experience. The user will find that their predictions are borne out in reality re-enforcing learning. An accurate mental model empowers users to solve problems which are novel and might not have been covered to any great extent in training. This can be a significant asset would it comes to controlling unforeseen hazards as the user is able to predict potential limitations and participate in troubleshooting. However, the way in which we conceptualise the system may or may not reflect its true nature; our conceptual model is fundamentally based on our beliefs rather than necessarily the facts [21].

When an individual's mental model differs significantly from the actual design, this can lead us to make the wrong decisions and to make them repeatedly. Designers in particular should not make assumptions about how users might picture the system, in some cases they may be the least qualified individuals to make that judgement as they are party to the technology's true architecture. Users often contaminate their mental model in ways we would not predict. These incorrect assumptions can originate from many diverse sources such as:

- Functionality and operation of legacy systems
- Behaviours observed in similar products in other healthcare organisations or departments
- Characteristics of other software products running on the same platform
- The media
- Hearsay from other users

Of particular interest are those assumptions which users make about a system based on their experience with other products and technologies. IT literate personnel may have already clocked up thousands of hours operating any number of other disparate applications, websites and mobile apps.

Users become familiar with certain customs. For example, an individual would probably expect a frame containing an X in the top corner to close the window. There is no law which states that this rule must be obeyed by software designers but over time users we have developed conscious and subconscious behavioural expectations of the user interface. Break those unwritten rules and users describe the system as clunky or unintuitive. These assumptions and expectations can be entrenched and hinder transition to an unfamiliar technology. Designers face the difficult dilemma of whether to revise the user interface to fit with a user's mental model or to improve their models by providing additional user training.

Just occasionally one may encounter a system which behaves in a way that is grossly unexpected and counter-intuitive. For example, suppose a system allows the user to record a diagnosis but before being stored in the patient record the user has to navigate to a completely different screen to confirm their input. No clinician would realistically predict the need to do this and even with training it is likely users would occasionally forget to make the confirmation. The system of course may be working as designed, there may be no clear fault to point at but the system behaviour doesn't meet user needs and is engineered in a way that the individual would simply not expect. Issues of this nature require an accurate safety assessment and option appraisal. Any decision to persist with the approach without the issue being addressed needs to be carefully justified against the risk.

5.5 Incident Reporting

The objective of risk management and a safety culture should be a proactive treatment of clinical risk and therefore minimisation of adverse incidents arising from HIT. Nevertheless this should not detract from the notion that one can learn a great deal from analysing incidents or near misses. A patient safety incident is "an event or circumstance which could have resulted, or did result, in unnecessary harm to a patient" [22]. The ongoing study of incident reports can help identify common types of problems [23] and incident reporting has been cited as one of the major steps in improving the safety of HIT [24].

By encouraging a culture of self-declaration, those who are closest to observing a near miss or harm are granted a means to express in their own words how a hazardous situation came to pass. The intention is to address preventable events whilst forging a sense of responsibility and preserving patient confidentiality. Incident reporting systems have been successfully implemented in many diverse industries such as aviation, nuclear power, petrochemical processing, steel production, and the military [25].

Safety experts in aviation often refer to the notion of a "Just Culture" – a way of safety thinking that promotes a questioning attitude, is resistant to complacency, is committed to excellence, and fosters both personal accountability and corporate self-regulation in safety matters [26, 27]. The concept builds on the idea of a "no blame culture" but better distinguishes between culpable and non-culpable acts. A Just Culture calls for errors to be reported in the knowledge that the information will be used for learning and future prevention rather than as a basis for punitive action.

Analysing incidents is now central to patient safety initiatives worldwide [28]. At a local level incident reporting and review can provide a fast turn-around of problem identification and risk mitigation acting as a basis for continual clinical improvement. At regional and national levels gathering this essential data facilitates the identification of trends and common themes which can be used to formulate alerts, drive future policy and establish best practice. To aid in this potentially mammoth data collection task, a range of formal reporting systems have been put in place such as:

- The UK's National Reporting and Learning System
- The Australian Incident Monitoring System
- The US MEDMARX database
- A number of programmes managed by US Patient Safety Organisations.

Consistency in reporting can be an issue and the AHRQ have created a number of common formats to bring an element of uniformity [29]. Work is ongoing to create a common format for HIT incidents in some care settings [30].

Underreporting means that incident data tells us little about the true frequencies of errors [31] and the US Institute of Medicine note that a lack of reporting data is hindering the development of safer HIT systems [32]. Nevertheless an analysis can still enlighten us about situations which had potentially been unforeseen and even shortcomings in the SMS itself. Medical device regulation generally requires manufacturers to actively facilitate the reporting of faults. Most developed countries have authorities in place to gather medical device incident data and bring sanctions against manufactures who fail to take action. But this is not the case for systems which fall outside of medical device regulation. In these cases healthcare organisations rely on voluntary clinical incident reporting and commercial arrangements with suppliers.

The role of incident reporting systems in HIT is still unclear from a practical perspective. Challenges remain in particular for ascertaining the precise role of technology in its contribution to an adverse incident, i.e. did it cause the event, fail to prevent it or simply remain a passive onlooker. Certainly the safety case should not be a replacement for sound incident reporting. But keeping a careful eye on local and national incident data will influence the hazards derived and the likelihood component of risk estimation.

5.6 Summary

- The hazards which HIT can introduce emerge from the complex interactions between technology and its human operators. Attempting to control risk exclusively in one or the other domain is naïve.
- A mature and informed approach to CRM can only be achieved through cultural shift. Whilst clear ownership and accountability are needed, blame is not a feature an effective method.

- Humans are inherently fallible especially in an environment where people and systems are competing for attention. Careful, user-centred design provides us with opportunities to reduce errors.
- The reporting of adverse incidents should not be relied upon as a sole means of identifying and managing potential hazards. However, the real-world performance of systems is important in assessing the effectiveness of controls and the on-going validity of the safety case.

References

1. Reason J. Human error. 1st ed. Cambridge: Cambridge University Press; 1990.
2. UK Department Of Health. An organisation with a memory. Report of an expert group on learning from adverse events in the NHS chaired by the Chief Medical Officer. 2000.
3. Sittig D, Singh H. Eight rights of safe electronic health record use. JAMA. 2009;302(10):1111–3.
4. Carayon P, Hundt A, Karsh B, Gurses A, Alvarado C, Smith M, et al. Work system design for patient safety: the SEIPS model. Qual Saf Health Care. 2006;15 Suppl 1:i50–8.
5. Peltu M, Eason K, Clegg C. How a sociotechnical approach can help NPfIT deliver better NHS patient care. [Online]. Cited 2015 June [http://www.bcs.org/upload/pdf/sociotechnical-approach-npfit.pdf]. Available from: HYPERLINK "http://www.bcs.org/upload/pdf/sociotechnical-approach-npfit.pdf" http://www.bcs.org/upload/pdf/sociotechnical-approach-npfit.pdf. 2008.
6. National Transportation Safety Board. NTSB Acting chairman says passenger ferries need an aggressive safety management system. NTSB Press Release. Office of Public Affairs. 2005 Nov.
7. Bayuk A. Aviation safety management systems as a template for aligning safety with business strategy in other industries. Creative Ventures International, LLC., 400 South 2nd Street, Suite 402–B, Philadelphia, PA 19147. Paper is here: http://www.scribd.com/doc/221557284/Safety-Mgmt#scribd.
8. Rasmussen J, Jenson A. Mental procedures in real-life tasks: a case study of electronic troubleshooting. Ergonomics. 1974;17:293–307.
9. Rasmussed J. Skills, rules, knowledge; signals, signs and symbols; and other distinctions in human performance models. Syst Man Cybern IEEE Trans. 1983;SMC-13(3):257–66.
10. Health and Safety Executive. Improving compliance with safety procedures. Reducing industrial violations. 1995.
11. International Organization for Standardization. IEC 62366:2007. Medical devices – Application of usability engineering to medical devices. 2007.
12. Lowry S, Quinn M, Ramaiah M, Schumacher R, Patterson E, North R, et al. Technical evaluation, testing, and validation of the usability of electronic health records. National Institute of Standards and Technology. U.S. Department of Commerce, Gaithersburg, Maryland. NISTIR 7804. 2012.
13. Gans D, Kralewski J, Hammons T, Dowd B. Medical groups' adoption of electronic health records and information systems. Health Aff. 2005;24(5):1323–33.
14. Belden J, Grayson R, Barnes J. Defining and testing EMR usability: principles and proposed methods of EMR usability evaluation and rating. 2009. Healthcare Information Management and Systems Society Electronic Health Record Usability Task Force. http://www.scribd.com/doc/221557284/Safety-Mgmt#scribd.
15. Microsoft Health. Common user interface. [Online]. 2015. Cited 2015 April. Available from: HYPERLINK "http://www.mscui.net/" http://www.mscui.net/.

16. National Patient Safety Agency. Design for patient safety. Guidelines for safe on-screen display of medication information. NHS Connecting For Health. 2010. London, UK. http://www.nrls.npsa.nhs.uk/resources/collections/design-for-patient-safety/?entryid45=66713.
17. Shneiderman B, Plaisant C, Cohen M, Jacobs S. Designing the user interface: strategies for effective human-computer interaction. Boston: Addison-Wesley; 2009. ISBN 0321537351; 2009.
18. Chandler P, Sweller J. Cognitive load theory and the format of instruction. Cogn Instr. 1991;8(4):293–332.
19. Tracy J, Albers M. Measuring cognitive load to test the usability of web sites. Usability and information design. 2006. http://www.researchgate.net/publication/253713707_Measuring_Cognitive_Load_to_Test_the_Usability_of_Web_Sites.
20. Shachak A, Hadas-Dayagi M, Ziv A, Reis S. Primary care physicians' use of an electronic medical record system: a cognitive task analysis. Gen Intern Med. 2009;24(3):341–8.
21. University College London Interaction Centre. CASSM project. [Online]. Available from: HYPERLINK "http://www.uclic.ucl.ac.uk/annb/CASSM/" http://www.uclic.ucl.ac.uk/annb/CASSM/. 2015.
22. Runciman W, Hibbert P, Thomson R, Van Der Schaaf T, Sherman H, Lewalle P. Towards an international classification for patient safety: key concepts and terms. Int J Qual Health Care. 2009;21(1):18–26.
23. Sparnon E, Marella W. The role of the electronic health record in patient safety events. Pa Patient Saf Advis. 2012;9(4):113–21.
24. Walker J, Carayon P, Leveson N, Paulus R, Tooker J, Chin H, et al. EHR safety: the way forward to safe and effective systems. J Am Med Inform Assoc. 2008;15(3):272–7.
25. Barach P, Small SD. Reporting and preventing medical mishaps: lessons from non-medical near miss reporting systems. BMJ. 2000 Mar 18;320(7237):759–763.
26. GAIN. A roadmap to a just culture: enhancing the safety environment. 2004. http://flightsafety.org/files/just_culture.pdf.
27. Dekker S. Just culture. 2nd ed. Aldershot: Ashgate Publishing Limited; 2012. ISBN 9781409440611.
28. Magrabi F, Ong M, Runciman W, Coiera E. An analysis of computer-related patient safety incidents to inform the development of a classification. J Am Med Inform Assoc. 2010;17(6):663–70.
29. Agency for Healthcare Research and Quality. Patient safety organisation programme. Common formats. [Online]. Available from: HYPERLINK "http://www.pso.ahrq.gov/common" http://www.pso.ahrq.gov/common. 2015.
30. O'Reilly. American Medical News. Better patient safety is goal of confidential EHR error reports. [Online]. Cited 2015 July. Available from: "http://www.amednews.com/article/20130724/profession/130729986/8/" http://www.amednews.com/article/20130724/profession/130729986/8/.2013.
31. Blandford A, Buchanan G, Curzon P, Furniss D, Thimbleby H. Who's looking? Invisible problems with interactive medical devices. 2010. http://www.chi-med.ac.uk/publicdocs/WP002.pdf.
32. Institute of Medicine. Health IT and patient safety: building safer systems for better care. The National Academies Press; 2012.

Chapter 6
Failure of Health IT

6.1 Dependable Computing

Computer systems are characterised by four fundamental properties: functionality, performance, cost and dependability [1]. Dependability is the quality of the delivered service such that reliance can justifiably be placed on this service [2]. Dependability is therefore related to the extent that we can trust the system to be able to support our needs and is generally considered to be a highly desirable characteristic of a HIT service. Dependability encompasses a number of attributes [3]:

- Availability – readiness for correct service.
- Reliability – continuity of correct service.
- Maintainability – ability to undergo modifications
- Safety – absence of catastrophic consequences on the user(s) and the environment
- Integrity – absence of improper system alterations
- Confidentiality – the absence of unauthorized disclosure of information.

Security is sometimes quoted as a further attribute whilst other texts consider it to be a composite of confidentiality, integrity and availability. Note that the terms availability and reliability are often confused and examination of their definitions highlights an important difference. Availability is how consistently we are able to gain access to the functionality we require when we need it. Reliability on the other hand is about preserving operation of the system once we have access to it. So, not being able to log on because the application server had been shut down would impact availability. The system terminating execution of a report prematurely would affect reliability.

A. Stavert-Dobson, *Health Information Systems: Managing Clinical Risk*,
Health Informatics, DOI 10.1007/978-3-319-26612-1_6

6.2 System Failure

Laprie [4] describes a number of threats to dependability and the means by which these may result in system failure. Failure occurs when the delivered service deviates from the agreed description of the expected service (assuming of course that the description is accurate). Failure usually originates from a fault (a defect in the system). The fault gives rise to one or more errors – a discrepancy between the intended behaviour of a system and its actual behaviour. Failure occurs when the activated error is manifest as an impact on the service.

Whether or not an error will lead to a failure depends on several factors which include:

- The conditions whereupon an error which is dormant becomes activated
- The system's composition and the amount of available redundancy
- The definition of failure from the user's perspective.

6.3 Causes and Failure Modes

'Cause' is a term frequently used in safety and risk management. Although regularly found in safety standards and the safety engineering literature it is rarely provided with a formal definition. The term usually relates to those factors which are credibly capable of triggering a hazard. It is the wide scope of the term, encompassing the service, its operators and its direct relationship to the hazard register which is appealing. This book uses the word extensively and its meaning remains independent of whether the underlying causal concept is a fault, defect, error, failure, slip, lapse, mistake or other similar terms.

The term 'root cause' is also found frequently in the literature and it is worth contrasting this with the use of the term 'cause'. For harm to occur there will typically be a chain of contributing factors and events which lead to an eventual outcome. The root cause is the most basic cause which we have reasonable influence over and is responsible for the problem. Logically then, managing or eliminating this cause will eliminate or significantly reduce the likelihood of the problem occurring. In practice a problem may have multiple root causes each of which contributes to the problem being triggered. Note that increasingly the terms 'initiating events' and 'contributing factors' are replacing the use of 'root cause'.

Root cause is intimately related to a systematic method called Root Cause Analysis (RCA). RCA attempts to identify the factors which contribute to an incident and then work back to determine the underlying causes with a view to addressing them. RCA is often employed in the investigation of large scale disasters such as an air crash or mining collapse. In HIT RCA can be helpful when examining complex live incidents where the system is implicated. The technique is used widely in healthcare as an error analysis tool [5] along with other modified techniques such as the London Protocol [6]. RCA is less useful when it comes to developing the hazard register during the pre-deployment phase of a project and other methods may be preferable for this purpose (see Sect. 13.2).

Another term 'failure mode' inherited from safety engineering is often used to denote an unintended condition of a system. Some would argue that failure mode and cause are synonymous but from time to time they are used in a manner which it subtly different. Consider a failure of the power supply to an application server. The failure modes might be set out as; no voltage, voltage too high, voltage too low. To some extent the term implies that each potential failure mode is discrete and that collectively all possible eventualities are addressed (as with this example).

An analysis of a system's failure modes may determine that some could trigger a hazard – these failure modes are 'safety-related' and are likely to represent causes. In contrast those which would not typically lead to harm will be 'not safety-related'. Depending on the relative degree of risk it may be necessary to justify the reason for a failure mode not being safety-related.

Note that 'Failure mode' has also been adopted outside of health in the naming of a commonly used formal method for identifying hazards, Failure Mode and Effects Analysis described in Sect. 13.6.1.

6.4 Classification of Causes

It is impossible in a text such as this to list out the potential causes of clinical risk in any given system; to attempt to do so would undermine the need for a carefully though-out, multidisciplinary analysis. However, a classification system for potential causes turns out to be surprisingly helpful. We will see in Chap. 13 that a method for deriving hazards and their causes is essential in demonstrating completeness of a hazard analysis. A framework on which to hang causes of clinical risk is a useful stimulus for driving out hazard register content. Similarly the formulation of cause categories can aid the process of incident management and in eliciting the relevant information from incident reporters. Note that not every classification necessarily captures the scope of all possible causes. In some cases it is necessary to adopt a combination of approaches.

- System versus human-system interface failure – At a basic level, the causes of clinical risk can be differentiated as primarily pertaining to either the technology or to the human who operates it. Magrabi et al. [7] found that 55 % of reported incidents were primarily caused by the system whereas 45 % were attributable to the user. Whilst classification in this way can be useful as an aide-memoire to think about both technical and human factor causes, the socio-technical model tells us that such a binary classification is overly simplistic (see Sect. 5.1).
- Software versus hardware failure – Hardware relates to all physical aspects of the system and its components. This includes the application and database servers, storage systems, network connections and power supplies as well as the local peripheral equipment such as printers, monitors, keyboards, mice, etc. All

man-made equipment has a 'mean time to failure', in other words all hardware will at some point fail it is just a question of when or how often. Software failure represents a defect in the way the system was programmed. Of course that says nothing about whether the defect relates to behaviour which was truly unintended (a bug), whether the design was incorrect or whether the requirements for the functionality in the first place were inaccurate.
- Classification by ownership – Occasionally it is helpful to categorise causes primarily by the stakeholder who is responsible for implementing its controls. For example, suppose a hazard has been identified in relation to system unavailability. The causes might include failure to maintain local hardware, actions taken by the manufacturer's service management team, networking issues, configuration changes, etc. Each of these causes fall within the domain of different teams and this can facilitate communicating the necessary controls across departmental or organisational boundaries.
- Entry/retrieval versus communication/co-ordination – Ash et al. [8] describes two categories, one relating to entering and retrieving information and the other to issues of communication and care co-ordination. Ash found that issues in entering and retrieving information were often a result of poor user interface design, confused patient identity, overly structured information and fragmentation of the record. In contrast problems with communication and co-ordination were related to failures in representing real-life clinical workflow, system inflexibility, failure of clinicians ensuring completeness of communication and poorly managed decision support.

An alternative (or additional) approach to cause classification is to consider the impact on the clinical data which resides within the system – after all it is ultimately the information and how it is used which gives rise to harm. By examining how information is generated and presented we can derive a series of failure modes (or at least categories of failure modes) which have the potential to impact care. Depending on how granular one decides to cast the hazards and causes, the classification system can be used to drive out both.

Such a classification framework is presented in the following sections. The categories are not based on current standards or wider literature but rather from the author's personal field experience. They are not necessarily mutually exclusive or even complete but they offer a starting point to be used alongside other methods.

6.5 Absent Clinical Data

A clinician may be of the belief that a piece of clinical data should exist in the system when, for whatever reason it does not or it cannot be accessed. Alternatively a clinician may assume that if some relevant data does exist that this will be brought to his/her attention in the course of interacting with the system. Both of these scenarios result in situations where information is unavailable to the clinician requiring them to make assumptions or source the data from elsewhere.

There are a number of underlying causes which could result in information being unavailable:

1. Complete service failure
2. Modular failure
3. Functional failure
4. Data delivery failure
5. Poor system performance
6. Inappropriate denial of access
7. Elusive data

6.5.1 Complete Service Failure

In complete service failure, the entire system is typically unavailable, a situation which is often called an outage. This type of unavailability may of course be either planned or unplanned and may be expected or unexpected by the individual user. Service failures are often due to catastrophic hardware failure where insufficient redundancy has been employed. The clinical risk is determined by a number of factors:

- The duration of the outage
- Whether it occurs at a time of day when many users require access to the system
- Whether it was planned or unplanned
- The extent and effectiveness of business continuity planning
- The nature of the clinical business processes supported
- The number of users who rely upon the service
- The immediacy with which users require information from the system

The last of these is particularly important in assessing the risk. For some systems the supported business processes are not urgent and short periods of downtime can easily be absorbed. In other cases there is a more immediate need and this can occasionally be a significant source of risk. For example, where healthcare organisations are reliant on an electronic system to schedule drug administration, very significant clinical risk arises if that system suddenly becomes unavailable. Indeed for a clinician managing short half-life drug infusions, it can feel as though the lights have gone out. These scenarios are predictable when technology is relied upon to this extent so resilience and robust business continuity plans need to be in place.

6.5.2 Module and Functional Failure

Modular failure is the unavailability of a discrete part of the system (for example, the theatre management functionality or appointment booking screens). Functional failure is the inability to perform a specific task (e.g. allocate a patient to a specific theatre schedule). Modular and functional failures are often the consequence of defects in the software, its interfaces or the on-going management of the service. For example, a system might be upgraded overnight only to find that a software bug now renders part of the system unavailable. The potential impact in these cases is more subtle and varied compared to a complete system outage and depends on the precise nature of the module or functionality in question.

6.5.3 Data Delivery Failure

Data delivery failure relates to any set of circumstances where data is known to be present in the system but, for whatever reason, the system fails to display it to the user. This could occur as a result of a screen rendering issue or the failure of a report to execute. Some situations can be very subtle and difficult to detect. For example, a system intended to chart vital signs could have a confusing scale, be too small to interpret or fail to display correctly on a mobile platform.

6.5.4 Poor System Performance

A system becomes non-performant when the speed at which it operates falls below what is specified in the requirements. The impact can be anything from a minor irritant to effective unavailability and can range from an occasional intermittent problem to a persistent characteristic of the system. When the problem is severe users may abandon the system completely and resort to alternative methods of information provision and document keeping. In these cases the impact becomes essentially the same as system unavailability.

A further discussion on poor system performance is provided in Sect. 7.5.

6.5.5 Inappropriate Denial of Access

Inappropriate denial of access (IDA) describes a situation where a user should have the ability to access a piece of information but the system denies them entry. IDA can be a frequent problem for newly implemented systems. The balance between information governance (i.e. the responsibility to protect information) and the need for the relevant clinicians to be able to view the data they require to deliver care is a difficult technical, practical and sometimes ethical issue. Applications will vary in their approach to permission setting. Some will grant or deny access to individual screens whilst others work at a functional level. Similarly some systems allow groups of users with a common business function to share permissions, so called role-based access.

IDA manifests itself in at least three forms:

- Failure to correctly allocate user permissions – a system administrator will typically allocate permissions to individual users as they are registered on the system. Failure to do that appropriately can give rise to IDA. Organisations should have policies and training in place to ensure that the default permissions are appropriate to an individual's business function (e.g. junior doctor, nurse, physiotherapist, administrator, etc.). Defining the policies generally requires close cooperation between system managers and users.
- Insufficient permission granularity – some systems manage permissions at an inappropriately high level. In this way the system administrator is unable to grant permission to a system function without also providing access to unwanted functions. This is essentially a design issue as the system administrator is unable to devise a permission set which meets both practical and information governance requirements.
- Incorrect implementation of user permissions – IDA could occur as a result of a system defect whereby access is prevented for a user whose permissions are set quite correctly.

Note that in cases of information unavailability either:

- The user is fully aware that the information should be present but for some reason it isn't (a known unknown), or
- The missing information passes them by such that they are unable to question its absence (an unknown unknown).

The difference between these scenarios is one of detectability which is discussed in Sect. 14.1.3.

6.5.6 *Elusive Data*

Occasionally one encounters the situation where information is present within a system but it remains difficult for the user to find. Most patient records contain a mixed bag of disparate facts, opinion and supposition and it is a challenge for system designers to provide users with an intuitive means of navigating and making sense of this rich data. The clinical record can be visualised as a series of pockets in which nuggets of information reside; perhaps one for medications, orders, appointments, admissions, etc. Some pockets will be deeper than others and not all will contain information relevant to every clinician. With potentially many hundreds of pockets to put our hands into, which ones do we choose for any particular patient encounter given the time constraints clinicians face?

To some extent it is the responsibility of the clinician to correctly delve into the pockets which are relevant to supporting the clinical decision they are faced with. However, there has to be some recognition that unless a pocket is seen to be bulging with data, it may not always be investigated. Designers have a responsibility to engineer systems with the intelligence to make users aware of relevant information (within information governance boundaries). That requires user engagement and frankly some careful educated guesswork given the breadth of the clinical record. On occasion the design can be suboptimal resulting in a system which holds critical information but fails to make users aware of its presence.

For example, suppose a system has functionality to maintain a record of patient allergies. It is common in many systems to represent the presence of an allergy using an icon or other indicator in a permanently visible banner bar (in addition to a formal set of allergy management screens). Suppose that indicator was defective and failed to indicate the presence of an allergy. Even though the allergies could be found should one take the active step of delving into the 'allergy' pocket, the lack of an indicator would convince us that doing so would be unfruitful. In this way, the information is present within the system but, from the perspective of the user's workflow, might as well be absent.

The risk associated with issues of this nature depends on the extent to which the hidden information is relied upon clinically. But there is another factor to consider; when information is inappropriately hidden not only do we fail to take it into account but from time to time clinicians will make positive assumptions based on its apparent absence. For example, suppose the system fails to correctly indicate that a CT scan has been carried out when in fact it has. Firstly, by not reviewing the result of the scan we may introduce a delay in care and secondly, we might fail to acquaint ourselves with any clinical activity triggered by the result of the CT scan. Finally we may assume that the original order has failed and issue a new request – unnecessarily exposing the patient to additional ionising radiation. Whilst these user behaviours can be difficult to predict they should at least be considered by the analytical team during hazard analysis.

6.6 Misleading Clinical Data

Misleading clinical data relates to information which is available to the clinician but its nature or presentation is such that it fails to convey the meaning intended by its original author. Data can be misleading by nature of its:

1. Identity
2. Context
3. Presentation
4. Input
5. Corruption
6. Miscommunication
7. Direction

6.6.1 Misleading Identity

Most clinical information within an HIT system pertains to a specific patient. It is a basic requirement of most systems that the presented information carries the necessary markers to relate it back to a particular patient. Failure by the user or the system to identify information can introduce significant risk. In many cases it can be difficult to detect misidentification as the information appears otherwise credible and convincing. Often the error can adversely impact two patients; the one to whom the information was mistakenly attributed and the one who was the information truly belonged – one is provided with inappropriate care whilst the other is potentially denied care.

As a guiding design principle, it should be possible to immediately identify the patient to whom information pertains in just a single glance at the display. It should not be necessary to close down or reposition windows to view identifying data. In particular, users should never be required to remember the patient they have selected for care especially given the propensity for clinical users to be regularly distracted by their environment. The patient context should in all cases be explicit. There remains a responsibility on users of course to ensure that the real-world patient they are working with does indeed match the identifying information on the screen.

Beyond the user interface, patient identity can be confused for many other reasons, for example:

- Creation and preservation of patient unique identifiers
- Inappropriate creation of multiple records for the same patient
- Input of information against the wrong patient
- Reconciliation of incoming messages with local records especially when messages are unsolicited
- The management of patients whose identity is genuinely unknown
- Compromised database keys or referential integrity issues
- Functionality and procedures for merging duplicated patient records
- Preservation of identity during data migration and database merging

Two identity problems appear to be particularly prevalent in HIT systems. Firstly preservation of identity becomes particularly complex when it is possible to have multiple instances of an application open at the same time. Should each of those instances spawn new independent windows it becomes almost impossible to reliably distinguish one patient's data from another. Finding a solution to this requires a combination of sound design decisions, consultation and local policy particularly as some users find that having multiple instances of an application open has significant usability and productivity benefits.

Secondly designers often go a long way to ensure that the system sufficiently identifies clinical data on the screen but pay less attention to the challenge of identifying printed information. It is not uncommon to see a full set of patient demographics printed on the first page of document but for subsequent pages to lack any identifying features at all. As such, should a page be filed in the wrong patient's paper notes this could significantly compromise care. Add to this the tendency for users to share printers and a very significant hazard slowly emerges. Designers should be aware that any sheet of paper which fails to contain identifying information should be destined for the shredder and nowhere else. Certainly identity should never be inferred from the proximity of one sheet of paper to another.

6.6.2 Misleading Context

It is a peculiarity of HIT systems that so much interpretive weight is placed on the context of clinical data. The same piece of information captured in a slightly different set of circumstances can significantly change its meaning and the conviction that is placed in its validity. For example consider a blood pressure measurement retrieved by the system. In the clinician's mind there will be approximate upper and lower limits for what might be considered normal. One might then provide clinical intervention should one or more readings be outside of those limits. However the context in which blood pressure measurements are taken can greatly influence what might be considered normal. An anxious patient will have a higher blood pressure than when they are relaxed, they will have a lower blood pressure when laying down, the reading might be impacted by medication they were taking at the time or other pathological conditions. Thus without capturing and subsequently viewing the full context of the measurement, assumptions can be made and decision making potentially compromised.

A similar situation exists when one considers how the interpretation of documented opinion varies depending on the role and status of the information author. Suppose the patient record contains the following nugget of information "I suspect this patient may be suffering from Rheumatoid Arthritis". The author of this data could be a General Practitioner, District Nurse, Physiotherapist, Rheumatology Consultant or any number of others. Rightly or wrongly a clinician viewing the statement is more likely to have faith in the diagnosis if the author is a specialist in that clinical area. Indeed the subsequent management of the patient may be completely different depending on the provenance of the opinion. In this way the context contributes as much to the interpretation as the data itself.

HIT systems are frequently designed to use and reuse information for different purposes. This has huge productivity advantages as users are relieved from having to re-key information each time it is needed. However, there is a tendency in some systems for information to be cherry-picked and be reused in a manner which fails to communicate the full context in which it was captured. Suppose an anxious pre-operative patient has a series of high blood pressure readings taken leading up to a surgical procedure. These could be blindly messaged out to a GP system giving the impression that the patient is chronically hypertensive and requires treatment. Suppose that a vague suspicion of Rheumatoid Arthritis by a junior member of staff is coded by the system as a firm diagnosis. Another clinician could easily misinterpret this impacting their clinical decision making.

A number of real-world examples are provided by case studies in the literature. The IoM highlight one particular case [9]:

> An abdominal ultrasound report in an electronic record appeared to indicate a blighted ovum, and a dilation and curettage (D&C) was performed a few days later. The patient returned to the ER [emergency room] 4 weeks after the D&C with abdominal pain. Repeat ultrasound revealed a 21-week pregnancy. A damaged fetus was delivered at 26 weeks. The ultrasound result had actually been obtained several weeks prior to the date of the record in which it appeared. The report had been copied forward into that record and appeared out of context.

By overly reusing and simplifying clinical data we have the potential to strip away the context in which it was originally captured and this can introduce significant risk when incorrect assumptions are made during interpretation.

6.6.3 Misleading Presentation

Whilst the underlying clinical data in an application may be representative of the author's intention it is quite possible for the system to fail at the final hurdle by presenting that data to the user in a manner which misleads a user.

One common reason for suboptimal display of clinical data is simply a weakness in the system design. This can occur for many reasons including constraints of the technology, insufficient engagement with users, a manufacturer's lack of understanding of the requirements, configuration issues, etc. It is not the intention of this book to set out all of the principles of good user-interface design however a little common sense applied by users familiar with the clinical environment can assist in steering the design in the right direction.

It can be useful for designers to draw up a list of best-practise examples of solid user interface design. These may subsequently form the basis of some user interface safety requirements. The list might include for example:

- Information which is critical to the safety of the patient should be readily visible without requiring the user to scroll the screen where at all possible. For example, it often makes sense to list a patient's allergies by order of decreasing severity. Should there be a significant number of allergies so that scrolling or paging is unavoidable, it would be the least severe allergies which would be obscured by default. Similarly, consideration should be given as to how free-text fields will display when their content is long and screen real-estate limited.
- Each piece of information should have its place in the system and duplication should be avoided. As systems evolve they often suffer from having multiple places where the same piece of information can be recorded. If those screens fail to communicate with each other, a user may consult only one area and make incorrect assumptions.
- Numerical information should always be supplemented by units of measure and the units should be displayed in a consistent and unambiguous way, for example 'micrograms' instead of 'mcg' (which could be mistaken for milligrams).

Many applications use a browser of some description on the client machine to access the service. There are a large number of different browsers available for example Microsoft Internet Explorer, Google Chrome, Mozilla, Firefox, etc. Whilst these applications all intend to fulfil the same purpose there are subtle differences between them which have the potential to render data differently. It is often difficult for developers to write code which produces identical output in all browsers. A further level of complexity is introduced as each of the potential browsers typically exists in a number of different versions – the precise manner in which data is rendered may vary between these versions. In some cases the inconsistencies in display can be very subtle, for example the way in which special characters such as < & and apostrophe are shown.

The result is that clinical information can be presented to a user in a way that was not predicted by the manufacturer. Indeed it is unlikely that a manufacturer will have been realistically able to foresee and test the application on all versions of all browsers making it difficult even to predict what presentation hazards might arise. Most manufacturers will issue specifications which include the system's supported browsers however few lock the application down to a particular browser and version by design. In an age where Bring Your Own Device (BYOD) is becoming a more prevalent deployment strategy, it can be a challenge for healthcare organisations to effectively govern the means by which a user accesses the system.

A similar situation exists when one considers the physical size and resolution of the screen on which the application is viewed. Historically most systems have been designed from the ground up with a particular screen resolution in mind. As computer monitors have become larger and less costly designers have been able to support higher resolutions and include increasing amounts of data on the screen therefore reducing the numbers of clicks needed to access the information one might need. This becomes a problem when those same screens are accessed in their native form from a mobile platform which is typically much smaller than its desktop counterpart. The results can yield microscopically sized text and screens which require constant scrolling and/or zooming. A designer may have assumed that a particular portion of the screen will always be visible (e.g. a banner bar) and use it to site information critical to the safe care of the patient. On a device with a small form factor viewing of the banner might require a number of swipes and its content could be easily missed.

Increasingly manufacturers are now augmenting desktop systems with the provision of a supporting mobile app. These typically allow operation of the system via a bespoke and carefully crafted user interface whilst still remaining consistent with the desktop version. However mobile platforms still vary significantly in their size and supported resolution so careful thought should be given to the potential hazards which might arise and the extent of testing required.

For hazards relating to presentation, the stakes are raised when novel or complex methods are used to re-present clinical data in a different form. Graphs and charts are often used in HIT to illustrate a timeline or a trended view of biological variables such as blood pressure or plasma haemoglobin concentration. However accurately and consistently representing this information in graphical form can be a challenge. Software suppliers sometimes use third-party charting components that were not originally designed for the display of clinical data and these can present the information suboptimally.

A discrete instance of a biological variable is far less useful than a view of the changes in that variable over time. For example, a blood pressure of 150/85 provides a clinician with a limited amount of information. In contrast showing that 150/85 is part of a steady rise in blood pressure each and every month for the past year tells a very different story. Thus, trended data is often relied upon to inform clinical care. It follows that misrepresentation of this data can be a very real hazard.

Challenges for graphing clinical data include:

- Units of measure – some biological variables (e.g. blood gas results) can be reported using different units of measure. Mixing these in a single chart or graph can be very misleading.
- Negative numbers – some reported values are in the negative range (e.g. base excess). Charting tools can fail to represent these negative numbers. In some cases they may display both negative and positive numbers as positive numbers which can be very misleading.
- Reference ranges – Investigation results are frequently reported alongside a 'normal range'. The range can change from one measurement to another (depending on the reagents used, temperature of the lab, etc.). This can be a challenging to convey on a chart. Some charting tools will only show a reference range for the biological variable as a whole rather than for individual measurements.
- Axis scales – Deciding on an axis scale which works for the parameter(s) being plotted can be complex. For example, some variables utilise a logarithmic scale. Where time is plotted on the x-axis, there is often the issue of being able (or otherwise) to zoom in and out of that scale. Failure to get this right can make a graph difficult to interpret.

In clinical noting, a picture often paints a thousand words. Clinicians routinely draw diagrams and pictures in paper notes as a convenient way to illustrate what they observe in real life. But this simple task can also have its challenges when it comes to translating this into electronic form. The solution frequently offered by HIT systems involves presenting the user with a series of anatomical template images. The user selects the relevant template then annotates it before saving to the record. Many systems allow the inclusion of text labels, the use of different colours or the placement of standard shapes such as arrows or dotted lines. Interestingly it is the richness of information depicted in these images which can lead to so much clinical confusion if the system fails to reproduce the annotator's original artwork.

For example, suppose a patient presents with pain in the right iliac fossa (lower right part of the abdomen) and this is documented in the electronic record as an annotated image. If the system were to shift the annotation (or the position of the template image) just very slightly, the marker might appear over the suprapubic region (lower middle part of the abdomen) or right loin. The small change could radically impact a clinician's thinking changing completely the differential diagnosis.

Rendering images correctly (template and annotation) is subject to many factors which could adversely impact their relative alignment. For example, the screen resolution, screen size, browser type, browser version, maximise/minimise window functionality and scrollbar operation. A further level of complexity emerges when one wants to change the template images and still preserve the data captured historically. If there is functionality to add text to the diagram, careful thought needs to be given to avoid overlap and how word-wrap is managed. None of these issues are impossible to solve but for many systems supporting this kind of functionality, hazards around annotated images containing misleading information need to be considered.

6.6.4 Misleading Input

However well designed an application might be, the quality of its output is highly dependent on the quality of the information input – the proverbial garbage in-garbage out. In an evaluation carried out by the ECRI Institute, 30 % of HIT related incidents were categorised as being related to the wrong input of data [10]. Similarly, in an analysis by Magrabi et al. [7] researchers found that human data input issues accounted for 17 % of reported incidents relating to HIT systems.

Of course whilst data input is ultimately the responsibility of the user it is naïve to assume that instructing users to be careful will prevent these kinds of mistakes. Often errors of this nature will have deeper underlying causes – a sub-optimally designed user interface, inadequate user training, environmental distractions, etc.

Once again the design of the user interface is often a key contributor to the potential causes of hazards in this area. Some examples of poor design might include:

- Requiring the user to select from long dropdown lists with the potential to click on a value neighbouring the one required
- The use of poorly described or ambiguous captions on data entry fields
- Routinely forcing the user to enter 'dummy data' into a mandatory field because the information is genuinely unknown
- Overly restrictive validation on input fields and inappropriately short character limits on free text boxes
- Badly configured reference data which results in the user having to select an option which inadequately reflects the real world clinical situation
- Inappropriate prompts, notifications or workflow guidance which directs the user to enter information in an incorrect way
- Insisting that users describe a clinical situation using text when an annotated image or other media would communicate a more accurate representation

Avoiding these pitfalls requires careful cooperation between designers and users along with the opportunity to re-evaluate the adequacy of the user interface once the system is fully configured and has been operated in a real-world care setting.

6.6.5 Data Corruption

Data corruption refers to unwanted changes in clinical information which were not knowingly initiated by a user. At the simplest level, information stored in the system database could be truncated or transformed into an unreadable format essentially resulting in a loss of information. Note that unlike presentation issues, corruption refers to fundamental changes in the underlying data itself independently of how that information is subsequently rendered in the user interface. In this sense corruption can be more serious as it may well be irreversible or at least require significant effort to restore.

Corruption of information can also occur whenever data is transferred from one logical or physical location to another – in HIT systems data transfer often takes the form of messaging in either a proprietary or standard format such as Health Language 7 (HL7). Data which is in transit between systems is often vulnerable and robust design measures are usually required to ensure that any corruption is detected and brought to the attention of an appropriate service manager.

In practice, corruption of clinical data is often more subtle than a simple scrambling of text. One potential failure mode which can give rise to corruption is the loss of referential integrity within the database. Most proprietary databases hold information in tables, collections of data with a common set of fields and semantic meaning in the real world. Information in one table is often linked to that in another and it is this linkage which is vulnerable.

For example, suppose the system receives a report of a full blood count investigation undertaken on the 1st May. The report itself is likely to consist of a number of individual components for example, plasma haemoglobin, white cell count, platelet count, etc. In the database the single 1st May report may be linked to each of the individual tests and assays which make up that report – this is called a one-to-many relationship. It is feasible that data corruption could result in a break in that linkage essentially orphaning the tests from the report itself. Alternatively the results could be associated with an old report or, worse still, to a report for a completely different patient. An error of this nature with potentially serious clinical consequences can be brought about simply by the corruption of a single index number in a table of data. Whilst it is unlikely that any specific intervention by the clinical user would cause such corruption, there are factors such as defects in the underlying database or errors in the code used to talk to the database which could initiate such a change.

Corruption of data is not something which generally happens spontaneously in most state of the art databases. Occasionally it is caused by tasks undertaken on the database itself. Databases require a certain amount of maintenance to keep them in good order. Database engines and their supporting infrastructure are often updated as are the applications which make use of the data. These factors sometimes require database administrators to move data or transform it in some way. These activities have the potential to impact large numbers of records and even small risks can become significant when the likelihood is multiplied across thousands or millions of data items. Indexes and keys which preserve referential integrity can be an occasional casualty in these bulk actions. For this reason, database management activities should be conducted by well-trained staff familiar with the system and necessary techniques and in a manner which is tested before it impacts the live environment.

A suitable variation on the theme of corruption occurs when data is deliberately or otherwise modified in such a way that information within the system becomes erroneously described. A common source of anguish in HIT is the management of reference data. Reference data constitutes those values which are permitted to be associated with other fields in the system, for example, the names of areas within a hospital, the different 'types' of admission or the local drug formulary. These datasets can change over time and form the bedrock of local configuration. Unlike other data though, one would not usually expect to see modification on a day to day basis. Reference data is easy to forget about, it sometimes goes for years without even being reviewed. Despite this, failure to manage these important datasets correctly can lead to some surprisingly serious hazards.

Take an example, imagine a new system administrator notices that, within the 'drug administration frequency' reference dataset, someone has accidentally created two data items '3 times a day' and 'three times a day'. He also notices that there is no entry for 'five times a day'. He can kill two birds with one stone; he accesses the redundant '3 times a day' option and replaces the text with 'five times a day'. Being a diligent individual he navigates to the prescribing screen and, sure enough, all options are appropriately present and both issues have been fixed.

What the individual has failed to appreciate is that many systems fail to apply any means of version control to reference data. Thus whilst all prescriptions from that point onwards will be accurate, when one observes historical prescriptions, every time the '3 times a day' option has been used the system may display 'five times a day'. This is clearly very misleading from a clinical perspective. In modifying one byte of data, the system administrator has potentially impacted thousands of prescriptions.

Managing reference data can be particularly troublesome when two or more systems are merged for example when two local systems are being replaced by a single alternative and there is a need to preserve historical clinical data. Each system is likely to have its own reference data and, particularly over time, it is common for these datasets to have similar purposes but with different content. For example, a Patient Administration System may have five different options for 'Admission type' whilst the Electronic Health Record may have seven different options for that same field. Whilst it might be possible to live with this discrepancy on a day-to-day basis, the situation suddenly becomes very complex when it is necessary to merge the two datasets. Any proposed solutions need a careful safety assessment.

Another level of complexity is introduced when standard datasets and vocabularies are implemented. National and international datasets such as SNOMED CT and ICD 10 are becoming increasingly common and bring with them enormous benefits of consistency and standardisation. However, their implementation is not straightforward and can occasionally be a source of clinical risk. For example:

- Is the dataset rich enough to cover all potential options?
- Can the dataset be appropriately navigated by those not intimately familiar with its structure?
- How does the introduction of a standard reference dataset impact information that is being maintained historically?
- How do datasets interrelate to each other? Is there a need for mapping?
- How will the dataset be updated when new versions become available?
- What will happen to terms in the dataset when there is a need to retire them or when existing terms need to be corrected?
- How should systems communicate when they use different vocabularies to mean the same thing?

The impact of reference data issues on the patient (and therefore the associated clinical risk) is very difficult to predict. Clearly it depends on the nature of the functionality that the reference data supports. What is important is that reference data is considered as part of every hazard assessment and that controls are developed. Typically this would include a reference data management policy and/or data migration strategy.

6.6.6 Miscommunication

Miscommunication relates to the failure to clearly and completely transfer information from one party to another and in the context of HIT there is one particular case which is of special importance. This is the situation whereby an individual experiences the illusion of communication when in fact the information was never received or actioned.

When communication occurs in real time, be that face to face, using telephony or electronic instant messaging it is clear to all parties that transmission and receipt of information has occurred. But when we interact asynchronously through systems such as email or application specific messaging we accept that there will probably be a delay until we hear from the recipient. The sender may or may not expect an acknowledgement and the receiver may or may not send one irrespective of whether they actually action the content of the message. In this way there are many opportunities for the sender to believe that they have achieved communication when this is not the case.

Many HIT business processes rely on the close co-operation of multiple individuals. Often a single individual with a duty of care will initiate a set of activities for a particular patient. For example, a clinician places an order for a full blood count or similar blood test. The request is perhaps sent to a local phlebotomist who obtains the sample and passes it to laboratory services. This is then processed, the results obtained and issued back to the individual (or their team) who requested the test in the first place. At each stage of the process, information may be recorded on the system to facilitate the hand-off to the next stage. Of course there are innumerable reasons why the process may not complete; the request may fail to be initiated, the phlebotomist may be sick, the sample may be insufficient, etc. Essentially the baton is dropped and the process discontinued or stalled.

The point is that providing the process originator is aware of the problem he/she is more than likely to be able to intervene and ensure that remedial action is taken. However if the individual with a duty of care remains unaware and unalerted to the issue they are likely to go about their business in blissful ignorance. The originator perceives that all is going to plan when in fact this is an illusion as a result of the system failing to actively report progress. Unfortunately it is often when the patient experiences harm that the electronic black hole suddenly becomes apparent.

This particular failure mode has the potential to impact any clinical business process where multiple individuals or systems are involved especially when there are natural delays in the process anyway. These processes might include referral management, confirmation and validation of prescriptions, appointments and scheduling, etc. In many cases these problems can be controlled through sound design and configuration providing they are foreseen. It could be considered good practice to provide users with summary screens on which they can view recent activity and detect unexpected delays. Alternatively systems may be configured to alert users when activities have breached time-bound thresholds. Whatever the solution, the key is to have a clearly identified task owner who at any time is able to determine the status of the clinical business processes in which they operate.

6.6.7 Misleading Direction

A number of modern HIT systems provide users not only with the means to record and review clinical information but to actively support the clinical decision making process itself. A decision support tool "is an active knowledge resource that uses patient data to generate case-specific advice which support decision making about individual patients by health professionals, the patients themselves or others concerned about them" [11].This functionality provides in-line advice at the time of executing a relevant workflow. For example, drug interaction checking when writing a prescription or suggesting a care pathways at the point of recording a diagnosis. The advice might be passive background information (such as general data about a selected drug) or active alerts based on the specific characteristics of the target patient. These systems have the potential to mislead should they offer information which is incorrect or fail to provide information on which the user is (or has become) reliant.

The European Commission's report eHealth for Safety summarises that "In the case of clinical and organisational decision support systems, internationally, there are over two decades of sound evidence on their benefits" [12]. Indeed, large numbers of healthcare organisations are turning to electronic decision support systems to forge consistency in care provision and develop best practice particularly in prescribing. But where clinicians become reliant on decision support to deliver safe clinical care, it is a short step to harm occurring when these systems are found to have shortcomings. The Institute of Medicine in their Building Safer Systems for Better Care notes that whilst prescribing decision support systems have benefits in reducing medication errors and in detecting toxic drug levels there is variation in the ability to detect drug-drug interactions and the potential for alert fatigue [9]. In some cases, the risk associated with automated decision making can make the product fall into medical device regulation.

The risk associated with hazards in this area is often a function of the extent to which the human user has the ability to apply clinical judgement and disregard the suggested direction. For example, where a system passively presents a prescriber with the known side effects of a drug, he/she is able to make an informed decision on the pharmacological pros and cons. But where a system behaves in an automatic manner, perhaps restricting access to certain drugs which the system believes to be inappropriate, the potential for user intervention (and therefore human factor controls) is reduced.

Hazards and issues relating to decision support include the potential for:

- The underlying decision support data to be incorrect or out of date such that the advice provided is misleading
- The decision support rules to be configured incorrectly so they are triggered at the wrong time or not at all
- The decision support engine itself to exhibit a fault such that the configured rules are applied incorrectly
- Clinicians to overly rely on the system and forge ahead without questioning the validity or applicability of the advice
- Staff to deskill in their decision making and/or become complacent in keeping their clinical knowledge up to date

Note that unlike many other causes of clinical risk, the failure modes in this category do not actually relate to the patient's clinical data at all. Instead it is the decision support content, rules, algorithms and associated functionality which trigger the hazard. Controls in this area are not just about assuring the system but also about make sure that the advice provided is of a suitable level of quality. In some cases the decision support content and rules might be provided by a third party and checks should be undertaken to ensure that the information is appropriately governed, commensurate with best practice and kept up to date.

6.7 Summary

- Clinicians needs systems which are dependable such that they can place trust in them. Dependability encompasses availability, reliability and maintainability.
- Causes and failure modes represent the variety of ways in which hazards can be triggered and they arise from many different sources.
- By having a framework it becomes easier to tease out potential hazards and causes in a HIT system.

- Clinical data can give rise to hazards when either it is not available to a clinician or if it is misleading in some way.
- Clinical data can be absent because all or part of the system is unavailable or slow, access to it is denied, its presence it not noticed or it fails to be delivered.
- Clinical data can be misleading because of how it is presented, contextualised, identified or entered. Additionally information can become corrupted, fail to be communicated or the system provide inaccurate advice such that decision making is compromised.

References

1. Avizienis A, Laprie C, Randell B. Fundamental concepts of dependability. Research Report No 1145. LAAS-CNRS. 2001.
2. Lardner D. Babbage's calculating engine. Edinburgh Review. Reprinted in P. Morrison and E. Morrison, editors. Charles Babbage and his calculating engines. Dover; 1961. 1834.
3. Avizienis A, Laprie J, Randell B, Landwehr C. Basic concepts and taxonomy of dependable and secure computing. IEEE Transacts Dependable Secure Comput. 2004;1(1):11–3.
4. Laprie J. Dependable computing and fault tolerance: concepts and terminology. Laboratory for analysis and architecture of systems. National Center for Scientific Research; 1985. http://www.macedo.ufba.br/conceptsANDTermonology.pdf.
5. Agency for Healthcare Research and Policy. Patient safety network. [Online]. 2015 [cited July 2015. Available from: http://www.psnet.ahrq.gov/primer.aspx?primerID=10.
6. Taylor-Adams S, Vincent C. Systems analysis of clinical incidents. The London Protocol. [Online]. [cited Jun 2015. Available from: http://www1.imperial.ac.uk/cpssq/cpssq_publications/resources_tools/the_london_protocol/.
7. Magrabi F, Ong M, Runciman W, Coiera E. An analysis of computer-related patient safety incidents to inform the development of a classification. J Am Med Inform Assoc. 2010;17(6):663–70.
8. Ash J, Berg M, Coiera E. Some Unintended Consequences of Information Technology in Health Care: The Nature of Patient Care Information System-related Errors. J Am Med Inform Assoc. 2004 March-April; 11(2):104–112.
9. Institute of Medicine. Health IT and patient safety: building safer systems for better care. Washington, DC: The National Academies Press; 2012.
10. ECRI Institute. ECRI Institute PSO deep dive: health information technology. ECRI Institute PSO; 2012.
11. Liu J, Wyatt J, Altman D. Decision tools in health care: focus on the problem, not the solution. BMC Med Inf Decis Mak. 2006;6(1):4. http://www.ncbi.nlm.nih.gov/pmc/articles/PMC1397808/.
12. Stroetmann V, Thierry J, Stroetmann K, Dobrev A. eHealth for Safety, Impact of ICT on patient safety and risk management; 2007. European Commission. Information Society and Media. https://ec.europa.eu/digital-agenda/en/news/ehealth-safety-impact-ict-patient-safety-and-risk-management-0.

Chapter 7
Availability and Performance of Health IT

Section 6.1 introduced the concept of dependable computing and the notion that availability and reliability are two of its key characteristics. A HIT system which is relied on clinically for the delivery of care can quickly result in a hazardous scenario when it becomes unavailable to users. This is particularly the case for systems that deliver data which is consumed quickly by its users and influences clinical decisions in a short timeframe. One study reported that for each minute of system downtime it took approximately 4.5 min for staff to complete the equivalent work had the system been operational [1].

The potential unavailability of a HIT system should always be anticipated. Whilst its timing may be unpredictable the fact that at some point access to the system will not be possible is virtually inevitable in view of the number of typical dependencies. Architectural and procedural measures can be employed to minimise the risk of unavailability but this should in no way translate into a guarantee of continuous operation. Healthcare organisations need to plan from the very start of a project how they will manage and cope without access to the system for short, medium and long periods.

Architectural measures of protecting the service often come at a price so balancing the risk with cost is an important consideration. For those developing controls and evidence for the safety case, a basic knowledge of the architectural features intended to provide resilience can be helpful. Certainly demonstrating that due consideration has been given to redundancy, fault tolerance, resilience, disaster recovery and business continuity is a more powerful argument than simply quoting a blunt commercial service level agreement.

Note that in discussing availability and performance one needs to make reference to the stakeholders responsible for hosting the system. In practice this could be the healthcare organisation, software manufacturer or a third party. The principles of fault tolerance and resilience remain the same irrespective of which party takes commercial responsibility. However the stakeholder responsible for implementing controls will vary as will the information available to safety case developers. In this

A. Stavert-Dobson, *Health Information Systems: Managing Clinical Risk*,
Health Informatics, DOI 10.1007/978-3-319-26612-1_7

text, the term 'hosting organisation' is used to generically refer to the commercial entity responsible for undertaking hosting in any particular scenario.

7.1 Planned Service Outages

Most systems require at least brief periods of planned downtime in order for technical activities to take place. This often occurs when system components are upgraded to new versions, as a result of hardware changes or alterations in architecture. Early in the lifecycle of a deployment, healthcare organisations will need to liaise closely with the manufacturer to determine the nature, frequency, timing and duration of typical planned outages.

Most organisations are able to continue to deliver care during a planned outage but a careful assessment must be taken on the risk that the period of unavailability could be extended should something unforeseen occur. For example, suppose there is a 12-h planned outage window to perform an extensive upgrade on an electronic records system. The transition design indicates that a rollback could be performed if the upgrade is unsuccessful. But, the rollback itself may take 6 h to complete. Thus the outage should be planned to last not 12 but 18 h. This might be the difference between strategically undertaking the activity on a weekday evening or over a weekend.

Thought should also be given to how the planned outage, its nature, scope and timing will be communicated to those potentially affected. For example, will some modules of the system still remain available, might the system be constrained to read-only access? Individuals need time to plan and put alternative business processes in place before the outage occurs. In large and disparate organisations notifying users alone can be a challenging task. Working out the logistics of mass communication takes time and to attempt to undertake this for the first time the day before an outage is somewhat foolhardy. A communication plan, developed in collaboration with users, should be drawn up long before go-live of the system. The extent to which groups of users are potentially impacted by the outage should be fully understood and documented in the hazard register. Once the first few planned outages have occurred it is likely that lessons will be learned and it may be necessary to update communication plans and indeed the safety material itself.

7.2 Unplanned Service Outages

From time to time most complex HIT systems unexpectedly become partly or fully unavailable. Modern systems are architected to minimise the impact of such occurrences on the level of service experienced by the user. Indeed some contracts penalise hosting organisations significantly should they arise. The move towards 'software as a service' means that systems can be hosted on highly resilient platforms in state of the art data centres with hosting organisations benefiting from economies of scale.

In Sect. 6.1 we established that dependability is a desirable characteristic of a HIT system but we also require systems to remain dependable in the face of threats and changes. This property is known as resilience from the Latin *resilire*, to rebound. Resilience can be defined as "the persistence of service delivery that can justifiably be trusted, when facing changes" [2]. A useful physical metaphor for resilience is to imagine a basin with a ball sitting in it. If a force is applied to the ball it will rattle around the basin for a while before settling back at the bottom, but push it too far and the ball will fall out of the basin altogether. The height of the basin's walls, and therefore its ability to contain the ball for a given force, is analogous to the system's resilience. In the context of HIT, a resilient system will for example continue to remain available and perform well during peaks of user activity, during single component failure or in the face of executing processor intensive tasks. In such circumstances a system which is not resilient may experience periods of unavailability, instability and/or slow performance.

In developing the requirements for a system healthcare organisations need to reflect on the period of time for which a service outage could occur without introducing unacceptable risk. This period is called the Recovery Time Objective or RTO. Note that the RTO is always defined by what is acceptable in the operational domain and not by the actual capabilities of a particular system or its manufacturer. It can be tempting to specify the RTO as zero but in the real world of HIT most care processes can continue at least to some extent in the face of system unavailability. Note however that the RTO may be different across care settings and consultation with users will be required to identify the most critically dependent services. The RTO is a useful benchmark to provide to manufacturers and hosting organisations as this is likely to influence the architectural design of the system and ultimately the cost.

7.3 Component Failure

In a complex datacentre component failure is a regular occurrence simply due to the sheer number of components involved and the fact that almost all components of a system are prone to one or more failure modes over their lifetime.

There should be an expectancy of failure in every component and this has been modelled with metrics such as Mean Time To Failure (MTTF) which is particularly useful for hardware components. MTTF is essentially the average life expectancy of a non-repairable item. Note though that this period is not finite but rather a mean – for every device which significantly outlives its predicted lifetime, another will suffer premature death. Sound engineering, quality design and the use of high specification sub-components improve reliability but no amount of effort will result in an immortal product. Instead a plan must be formulated by the hosting organisation to ensure that when individual components fail replacement parts are available and can be installed in a timely manner.

7.3.1 Redundancy

It does not necessarily follow that component failure automatically results in prolonged systemic unavailability. So long as we architect systems in a manner which is fault tolerant, service affecting incidents can be minimised. With this in mind a key approach to maintaining service availability is to identify those components which are architecturally single points of failure. This might be a single application server, database or disc drive or perhaps a more subtle element, the network's DHCP server or an individual router. Once those points of failure are known it makes sense that they should be the focus of mitigation strategies. The most common approach employed is that of redundancy.

Redundancy is a term often confused with resilience. In effect, redundancy is a strategy one may choose to implement in pursuit of a resilient system. It can be defined as the "Provision of multiple components or mechanisms to achieve the same function such that failure of one or more of the components or mechanisms does not prevent the performance of the function" [3]. In other words redundancy represents the inclusion of additional hardware or functionality which might not be critically required during normal operation but is seamlessly available should it be called upon in situations of failure. For example, one might choose to operate a system with not one but two databases hosted in different geographical locations and synchronised in real time. Should one fail the system can be designed to automatically revert to the alternative without any obvious impact on the user.

Whilst the provision of redundancy has no impact on the likelihood that individual components will fail, it provides a means by which the service can continue in an interrupted manner in the face of component failure. This gives personnel time to address the problem without the added pressure of dealing with an outage.

Components which commonly employ redundancy include:

- Power supplies – If there is an outage of local power an alternative may be provided so that service can continue. The substitute technology is not dependent on mains supply and may involve a mechanical generator or battery. It is also common for servers to be fed from two separate power supply units which are themselves fed from separate mains distribution circuits.
- Fans and cooling systems – Large servers produce considerable quantities of heat. If all of the server technology is placed in a single room, failure of cooling could cause several components to fail simultaneously. Some datacentres have duplicate, independent cooling systems to mitigate this problem.
- Hard disc drives – Most hard drives have moving parts and tend to fail more often than other components. In some large servers there may be hundreds or even thousands of individual hard drives which require almost constant replacement. Most systems are architected such that failure of one or more drives can occur without loss of data.
- Storage area networks (SANs) – These represent large numbers of hard drives which act as a single virtual data storage area. Although there is usually

redundancy within the SAN occasionally entire units can fail and may therefore be architecturally duplicated.

- Caches and short term storage – Many systems make use of temporary data storage areas which are low in capacity but fast in response. Failure of these might not impact long term storage but loss of cache data will almost certainly impact live user sessions. Some systems provide redundancy to maintain data in this area in the event of a failure.
- Servers – Individual servers can fail leaving the system devoid of any processing power. Resilient architectures employ multiple servers which can be seamlessly called upon when they are required. Increasingly this is achieved through a process called virtualisation where many servers exist logically within a single physical machine. The virtual servers can easily and quickly be brought on-line and with minimal effort.
- Local Area Networks (LANs) – The components of a system are usually connected together using a local network which itself has the potential to fail. The defect could relate to the physical cables, routers or other technology such as a DHCP server. Resilient systems often employ multiple network paths between components so that individual failures can be bypassed.
- Wide Area Networks (WANs) – No amount of redundancy within the data centre will protect the service if someone conducting highway maintenance nearby puts a spade through the network cable. Resilient systems often have multiple outward-facing network connections travelling in different geographical directions.

Provision of hardware or physical redundancy of course is not without its own challenges and one significant limitation is cost. Duplication of components will at the very least double hardware costs and increase requirements for more physical space, connectivity, power and cooling. Additional virtual servers on the other hand may not incur extra hardware costs. However, in both cases there may be additional charges for licenses and support. Similarly more technology inevitably introduces complexity and the process of fault finding can become increasingly cryptic. Finally of course not all software is capable of making use of redundant hardware even if it is provided. In fact in some cases the system itself may fail when redundancy is introduced should it not be supported fully. A careful balance needs to be struck between the clinical risk that system unavailability introduces against the cost and technical limitations of a resilient solution.

Redundancy is occasionally used in a slightly different manner when it comes to critical decision making. Suppose a system is responsible for the provision of advice based on a number of input parameters. Should the system fail the information provided could be incorrect or misleading and the error might not be detected by the user. One solution to this is for the decision making to be undertaken by not one but several independent components. Each component may be running identical or deliberately diverse code. Rather than any individual component making a unilateral decision, each one electronically votes for what it believes the correct output should be. One final component examines the votes and the winning advice is reported to the user. Where individual components recommend spurious actions, the divergence can be examined by service management personnel without impacting the timely delivery of a result to the user.

7.3.2 *Diversity*

One fundamental limitation of redundancy as a control is that some failure modes can impact both the active and redundant components. To address this problem requires technical diversity. Diversity can be defined as "A form of Redundancy in which the redundant elements use different (diverse) components, technologies or methods to reduce the probability that all of the elements fail simultaneously due to a common cause" [3]. Diversity is essentially a strategy for reducing the likelihood that redundant components themselves have a single point of failure. For example, suppose a system possesses a defect within the operating system. No amount of hardware redundancy will eliminate this problem as all instances of the operating system will be identical. An alternative strategy might be to provide an additional means of accessing the system (for example a 'safe mode' or even a different operating system) which might allow the defective code to be bypassed.

Techniques such as redundancy and diversity provide relatively strong engineering based controls for unavailability hazards. However the required architecture can introduce complexity and certainly cost. The relative risks need to be weighed against competing factors and a well-considered strategy involves engagement with organisations skilled in high system availability hosting services.

7.3.3 *Service Monitoring*

The management of component failure can be improved if one is able to quickly detect failure or impending failure. In fact one of the paradoxes of redundancy is that a malfunctioning component in a complex system can potentially go unnoticed as alternative components seamlessly take over. Many hosting providers choose to proactively monitor a service and its components not just to ascertain its availability but to proactively seek out deviant behaviour which might be a sign of impending failure. By constantly analysing system performance one might be able to detect deterioration and proactively take measures to remedy the situation prior to users detecting a problem.

For example, message queues are common candidates for monitoring activities. Incoming and outgoing messages are usually processed in turn and placed in some kind of buffer until they can be dealt with. Should downstream systems begin to under-perform messages will build up in the queue and detection of this can provide early warning that all is not well. Some software is specifically designed to report its own performance and well-being. Engineers will normally select critical or demanding functions for instrumentation and provide performance notifications either constantly or if pre-configured thresholds are breached.

Some hosting providers implement specific monitoring technology or operational support systems to capture system events, notifications and performance metrics. Pre-configured rules display this information to service management personnel who act in accordance with agreed policies when problems are detected. In some cases it may even be possible for operators to temporarily initiate service protection measures – changes in configuration which render the service in a more stable state but perhaps with limited functionality or with performance constraints.

Over time service management operators often develop a library of contingency measures which can be applied singularly or in combination to the system as difficulties arise. These processes introduce a degree of adaptation essentially allowing the system to flex as conditions change or threats arise. These valuable assets and the skilled operators who know how to work them act as risk controls in a very real sense. However, as with most things in HIT, implementing changes in the live environment can introduce risk in its own right. Bringing additional processors online with the intention of improving performance has the potential to crash the system or adversely impact other software executing on the same server. A data fix to correct an aberrant database record could adversely influence referential integrity with unpredictable results. Nevertheless, a tried and tested procedure, carefully executed under appropriate conditions has the potential to predictably rescue a failing service long before users become aware of the problem.

7.4 Catastrophic Failure scenarios

As architecturally resilient a system may be, redundancy at component level will not protect the system from catastrophic failure. For example a fire, earthquake or terrorist attack could quickly render an entire data centre useless. Whilst these scenarios are less frequent than component failure the stakes are higher in that recovery from these situations is challenging and the potential for permanent data loss a genuine possibility. A common solution to this is the use of a distributed architecture to achieve geographical diversity.

7.4.1 Distributed Architecture

Even at the level of catastrophic failure redundancy plays role. Many organisations choose to host their systems in not one but two or more data centres in geographically disparate locations. The chances of both data centres being affected by a single causative factor is very unlikely indeed. However this architecture introduces some new challenges. For example, designers need to consider how the two data centres will maintain synchronisation such that the active site can be seamlessly swapped to. Similarly difficulties can be faced when the application code or operating system needs to be upgraded at not one but two sites at the same time. Careful thought also needs to be given to the design of the network connecting the data centres together. This is likely to carry a significant amount of traffic to keep the two systems synchronised and it is important that the performance of the active site is not affected should this connection become compromised.

In a minority of cases systems are deliberately engineered to have a truly distributed architecture. These services operate across globally disparate physical hardware and provide very high levels of redundancy. Only a minority of applications make use of this technology and generally software has to be specifically crafted to operate in this kind of environment.

7.4.2 Disaster Recovery

Disaster recovery consists of the policies and procedures put in place to restore the service in the event of total failure. The term is often confused with business continuity which, in the case of HIT, represents continuation of care delivery rather than necessarily the technology which supports it.

In the event of actual catastrophic failure one's priorities turn to service reinstatement and in particular the restoration of data. Whilst unavailability of the service may present some short-term risk, permanent loss of current and historical clinical data could affect patient care for months or years. Organisations need to have a strategy for how data is backed up and restored and, in particular, the relative timings of those activities.

Many systems are backed up intermittently to high capacity removable drives dedicated to this purpose. The drive may then be physically transported to an archive which is geographically distant from the primary data centre. This process of backup and transport takes time and effort and the fact that it is undertaken intermittently means that the snapshot of data will be out of date the moment it is created. Should restoration of data be required it is only ever possible to reinstate the system to a point in the past. Transactions which have taken place since the backup was created run the risk of being irretrievable. The term 'data synchronisation point' represents the absolute point in time when the data snapshot was taken. After restoration of the system the state of affairs would resemble that at the synchronisation point. The further in the past the synchronisation point, the more data is permanently lost.

Some organisations choose to transport the backed up data electronically into a cloud storage system. This method makes regular backups more practical but even then extracting and transmitting the data takes time such that transactional gaps are still likely. In developing the safety case, healthcare organisations need to consider what would be the maximum time period for which transactions could be irretrievably lost without introducing significant risk. This time period is called the Recovery Point Objective or RPO. Knowledge of the RPO is likely to influence the backup strategy and should the RPO not be achievable in practice it could prove difficult to develop effective controls for hazards relating to catastrophic failure.

Should the criticality of the system be such that the RPO needs to effectively be zero, it may be determined that an intermittent backup strategy is simply inadequate. In these cases it may be possible for the system to be architected in such a way that critical storage components are mirrored onto geographically disparate servers. In this way whilst there may still be a delay in practically restoring the service, all data will still be intact right up to the point where failure occurred. Nonetheless even in these circumstances it is appropriate to take intermittent backups of the live data. For example, suppose a piece of defective code deletes critical data. The synchronised transaction will occur on both the live and mirrored sites potentially resulting in irretrievable data loss. Live data mirroring is therefore an adjunct to the backup strategy but not the complete solution.

One final factor is the testing of restoration activities. Whilst backup of data is something that is likely to occur every day, restoring the backup might only ever occur on a single occasion. The skills required to undertake this critically important activity may not be readily available and the policies and procedures surrounding the technique can easily go untested or, in many cases, simply be lost. In addition there can be practical difficulties in testing the restoration function. One couldn't routinely practise this in the live environment as it would adversely affects the service and test environments tend not to contain representative volumes of data. Those hosting critically important systems therefore need to develop solutions which ensure that personnel are confident and practised in the art of data restoration.

7.5 The Poorly Performing System

Performance relates to the rate at which systems are capable of processing data and delivering the required output. The performance of a system at any given time is determined by both the demands placed on the system and its inherent processing power. Should demand exceed supply then the system will develop a backlog of transactions and performance will typically fall. Should performance dip below the required level, system functions could be compromised and the user experience impacted.

The symptoms of poor performance can be varied but perhaps the most common (and frustrating) is a perceived lack of responsiveness at the user interface. This is particularly common when the processing power of the system is fixed and peaks in demand outstrip the system's capabilities. The service can be impacted in its entirety or slow responsiveness could be observed just on those screens which place significant demands on the technology. Where systems are co-hosted, high demand from one application can even affect the performance of another which is otherwise unrelated.

The routes by which harm could occur are varied but include scenarios where staff waste time waiting for the system to respond and the occasional tendency of users to abandon the system completely.

7.5.1 Time Wasting

Periods spent watching an egg-timer or a slowly moving progress bar is time one might otherwise spend providing clinical care. Waiting 30 s or so for a page to load can be frustrating but this wasted time can quickly add up when the poorly responding function is one that is repeated many times a day. Multiply this amongst the impacted users and a department or organisation can quickly become inefficient. The specific nature of the harm that could occur as a result is difficult to predict but this might not be insubstantial in a fast moving clinical environment. A hazard is also introduced when, out of frustration or boredom, users try to multitask whilst they wait for the system to respond. The resulting interruption in workflow and need to remember information between applications can be highly confusing.

It should however be noted that the actual time one is spent waiting for the system to respond is often much shorter than that perceived by a frustrated user. Waiting a few seconds might not be what we expect from modern technology but in assessing its association with clinical safety one must take an objective view on the true extent of the delay.

7.5.2 System Abandonment

A slowly performing user interface is certainly an irritant but there comes a point where the time one is required to wait for information simply cannot be justified. This is the threshold where the user effectively throws in the towel and chooses to provide care deprived of the requested data or with a need to source it from elsewhere. The extent to which one is prepared to tolerate a slowly performing system varies from one situation to the next and is at least in part dependent upon:

- The extent of the slow responsiveness
- The degree to which the clinician is reliant on the information required
- The perceived risk of continuing without the information
- The time pressures placed on the individual
- Social and psychological factors

A slowly performing application is likely to induce lack of trust in the system as well as a negative impression of the implementation project and its reputation. There are possible implications for the subsequent roll out of the system and the ability to derive benefits from it. All of these are very real problems which healthcare organisations will need to work through with their suppliers but ultimately these effects do not compromise patient safety. In characterising the clinical impact of a slowly responding user interface stakeholders should ensure that this is strictly defined in terms of the potential effect on the patient. It should be possible to generate simple functional examples of clinical activity which cannot be completed on time along with a description of the potential consequences for the patient.

Contaminating this sound safety argument with expressions of frustration, failure to meet project milestones and product reputation can lessen the safety argument's impact.

It should also be remembered that systems can experience slow performance in components which do not have an obvious user interface. Most solutions are architecturally modular and whilst the user interface may be perfectly responsive, a different part of the system may be performing sub-optimally. For example, inbound and outbound messaging can quickly be compromised where transactional demand outstrips processing power. The effect might not be obvious until a user notices that messages fail to arrive in a timely manner. Where the delay is beyond what is required and acceptable in the clinical environment, there is the potential for harm to occur.

7.6 Avoiding and Managing Slow Performance

The performance of a system is largely dependent on balancing transactional demand with processing power and communication bandwidth. One is therefore required to think about the specification of both in assessing likely performance. Note that most systems consist of large numbers of components and not all will be equally performant. The system as a whole will be compromised by the slowest dependency and determining which element that will be can be complex.

7.6.1 Factors Affecting Transactional Demand

Number of Users

Perhaps the most significant performance-impacting factor for most HIT systems is the number of users operating the system. Note however that there is an important distinction here between the total number of users with an account on the system and the number who are concurrently using the system at a point in time. The existence of a user account is likely to occupy no more than a few kilobytes of data in the database with little or no impact on overall system performance. However the number of users actively generating system transaction at any one time has a strong correlation with processor demand.

Systems can often be designed and scaled to meet the requirements of individual organisations. To mitigate the risk of poor system performance healthcare organisations should take appropriate measures at design time to estimate the numbers of concurrent users. Hosting organisations can use this information to predict the specification and architecture of system components required to meet performance requirements.

In many implementations healthcare organisations choose to roll out a system in stages gradually increasing the numbers of concurrent users as different locations and care settings join the project. Effort should be invested in predicting the rise of concurrent users over time at the start of a project. This valuable information can be used by manufacturers and hosting organisations to scale hardware gradually to support rising demand. High performance hardware is often expensive so developing estimates of user concurrency early in the lifecycle of a project facilitates the management of costs avoiding the situation where performance falls and funds are in short supply.

Code Efficiency

Use of a system can trigger many thousands of lines of computer code to execute but from a user perspective we merely observe the inputs and outputs of those functions. A programmer will tell you that there are very many ways of processing the same data to come up with exactly the same result. The efficiency with which the data is processed is dependent on the precise logic employed by the developer. Creating efficient data processing algorithms is no easy task. Those inexperienced in the technique can author routines which are perfectly functional but utilise mathematically redundant steps. Experienced programmers are often well versed in best practice occasionally reducing pages of complex code into a handful of carefully constructed operations.

For example, suppose one is tasked with creating an algorithm to search through a database of patient names to find those matching a particular patient. If the user wishes to search for "Mr Smith" one methodology would be to start at the first patient, compare their name to that being searched for and then move onto the next pulling out those with a positive match. The algorithm would only complete when the entire database had been successfully searched. Alternatively one could proactively create a 'key' against the appropriate database field. This time the database would know that names beginning with 's' don't start until the 150,000th patient and move on to 't' by the 160,000th. Suddenly the number of records that are required to be searched can be completed in a fraction of the time.

Failure to use efficient code can result in data processing bottlenecks particularly where the function in question is called frequently. Not only is the experience slow to the user but often whilst the system is processing inefficient code, it is not capable of doing anything else impacting other users of the system. What is key is that the problems associated with inefficient code can often be solved not by the upgrade of expensive hardware but by undertaking the relatively low cost task of tuning and improving processing logic.

Scheduling Processor Intensive Tasks

Some functions are likely to be significantly more processor-hungry than others. This is particularly the case for operations which are dependent on managing large amounts of data, for example:

- Execution of complex reports
- Bulk operations involving large numbers of patient records
- Database back-up and system housekeeping routines
- Migration of data from one area to another
- Installation or upgrade of system components

Many HIT systems experience peaks and troughs of user activity with the greatest transactional demand during office hours and a significant fall off during the early hours of the morning. Some systems exploit this variation by providing functionality to schedule processor-hungry tasks during periods of low user activity. This ensures that sufficient processing power is made available to these tasks without impacting the user experience. Clearly this is only possible for routines which are not critical for immediate patient care; but for selected jobs, scheduling intensive activity at night can significantly smooth the peaks and troughs of processing demands. Demonstrating that strategies such as these have been examined, developed and configured can be used as evidence in the safety case.

One potential hazard to be aware of is the possibility that a scheduled task could fail to complete during the low-activity window. Overruns have the potential to continue into periods of normal operation compromising system performance. This is a good example of a control introducing a new potential hazard.

7.6.2 Factors Affecting Processing Power and Bandwidth

Architecture and Scalability

Scalability is a property of a system that allows its processing power to be increased or decreased depending on the service requirements. The term usually implies that those changes can be effected without the need for complex modifications in architecture or code redevelopment. For example, many systems operate an architecture whereby the task of managing the application is distributed across several servers. A scalable system might allow more application servers to be added according to the expected demand (so-called horizontal scaling). This allows organisations hosting applications to price their services based on the customer's performance requirements. In contrast, a non-scalable solution would have an upper limit on its processing power and therefore the numbers of concurrent users it could support.

Some systems can be scaled vertically meaning that their processing power can be increased by adding more Central Processing Units (CPUs) and other fundamental components responsible for crunching data. This is often achieved through virtualisation where the system resides within a powerful computer often hosting many systems at the same time. Processing power can be allocated to specific applications as demand changes without the need to install new equipment.

The combination of carefully projected user numbers, well established application demands and a scalable architecture provides strong evidence to mitigate hazards relating to poor system performance.

Network and Local Hardware Specification

Occasionally it is not the power of central data processing which limits system performance but rather the capabilities of local infrastructure. The application itself might be hosted in a high performing, state of the art data centre but if the network connecting it to local workstations has limited bandwidth no amount of server processing power will overcome the server's restrictions. Magrabi et al. [4] found that 11 % of clinical HIT incidents reported were associated with failure of peripheral devices whilst a further 9 % related to slow performance or failure of a single workstation. Whilst the service itself remains operational in these circumstances, to the user the impact on clinical workflow is similar to general unavailability. Operational continuity will largely depend on access to alternative equipment.

Determining the required network bandwidth and architecture is a skilled and complex task. Healthcare organisations often choose to work with reputable partners to install wired or wireless network infrastructure specific to their requirements. The analysis and design activities which take place around this provide good evidence for performance related controls.

Similarly the specification of client workstations can also impact the perceived performance at the user interface. In the trend to move to cloud-based architectures the amount of processing undertaken by the client machine is often quite small. However this cannot be assumed and it important that healthcare organisations ensure that local technology meets the minimum specification provided by the manufacturer. Organisations should take into account the fact that client machines could be simultaneously running other software which could compete for processing power.

7.7 Business Continuity Planning for Healthcare Organisations

Unfortunately it is rarely possible to discontinue care delivery during a planned or unplanned outage, as they say, the show must go on. For this reason healthcare organisations should develop business continuity plans to decide in advance how they will cope. By formulating and documenting a Business Continuity Plan one is able to prepare for, agree and communicate the actions that should be taken by personnel in the event of an outage. The actions in the plan represent controls to unavailability hazards demonstrating that a considered and rational approach is in place to manage the clinical risk.

In general there are three areas which must be addressed in the plan:

- Unavailability of existing clinical data
- The inability to capture new clinical data
- Limitations in the ability to communicate clinical data in a timely manner

7.7.1 Unavailability of Existing Clinical Data

The first handicap encountered during an outage is generally the lack of an ability to access the data that is contained within the system. Without this information, clinicians are forced to make decisions devoid of key clinical facts. This might include historical notes, investigation results, previous admissions, etc. The impact of the outage will vary depending on:

(a) The extent to which the system is relied upon to deliver care, and
(b) The availability of alternative sources of information.

Clinicians may in some circumstances be able to delay decision making until the relevant information is made available. However, more immediate matters have the potential to be compromised. Thus it can be helpful in developing the Business Continuity Plan to establish which key data items are critical to care delivery and which are less significant. The resulting dataset is a good starting point for ascertaining potential technical solutions to managing the outage.

For example, one might establish that unavailability of the drug administration record and access to patient allergies would be the most immediate risk. One strategy might be to regularly take an electronic snapshot of this information for current inpatients at key points during the day. This could be maintained in a manner that does not rely on the same technology as the main application and therefore be made available during an outage. Although this information will always be out of date to some extent, with the application of clinical judgement it may be better than nothing during an outage.

7.7.2 Capture of New Clinical Data

Assuming clinical activity continues to be delivered during an outage thought must be given as to how data will be captured whilst the system is unavailable and how it will be retrospectively entered into the system. Without planning for this, patients will have a hole in their clinical record and clinicians will be left to make assumptions about what activity took place during this period. The effort required to input data retrospectively however might cause one to consider whether it is realistic to enter a complete dataset. Instead it may be appropriate to identify key data items for capture on paper or other systems and for these alone to be inputted on service resumption.

Pro-active analysis is also required to determine the impact of entering clinical data into a system retrospectively especially where business functions are dependent on the relative time that data was captured. For example, suppose a number of individuals are involved in an electronic dialog to carry out a blood test. The investigation is ordered, perhaps approved, the sample is taken, transported, analysed, verified and reported. Each of those stages may be represented in the database as a discrete task each with its own timestamp. Now should that continue during an outage using a paper system, how will that practically be entered into the system retrospectively? If on system restoration each actor records their contribution, the events may become disordered and chaotic with the potential to upset other workflow, business reports and newly spawned tasks. At a minimum systems need to have functionality to retrospectively capture the true timing of events and disassociate this from the time of data entry.

7.7.3 Communication of Critical Clinical Data

A significant amount of clinical data held in systems is captured only to be retrieved in a number of weeks or months. An outage in between time has little or no effect on the flow of this information. In contrast, some clinical business processes are associated with fast moving data in order to maintain continuity of care. For example, the ordering of tests and reporting of results, monitoring of vital signs and real-time tracking of patient location. Here long-term storage of this information is less significant, it is the use of the system as a communication tool which is key to operations – simple paper-based data capture will not meet the needs of a fast-paced clinical environment in this case.

Healthcare organisations need to recognise business processes which rely on the HIT system for real-time (or near real-time) communication. These areas will require specialist support during outage periods, in some cases employing physical measures to move paper from one location to the next. Use of a telephone as a back-up measure may be a reasonably sound approach but this is only useful if this is supplemented with supporting directory services and adequate provision of telephones. Alternatively it may be possible to continue business using other HIT systems which are functionally diverse from the main application. For example, support from instant messaging and email may be appropriate until the primary service is resumed.

7.8 Summary

- Service outages can be a significant source of risk particularly when systems are relied upon to deliver information which is used to support immediate care.
- System unavailability should be expected and planned for as no service is 100 % reliable.
- Healthcare organisations need to have business continuity plans in place and hosting organisations need tried and tested disaster recovery strategies. The existence of these plans and policies can be used to support the safety case.
- Sound architectural design is essential in eliminating single points of failure and in providing an appropriate level of redundancy.

- The technology supporting high reliability services is mature but this comes at a price. A balance needs to be struck between risk and cost.
- Users find slow performance of systems frustrating and in some cases clinicians may abandon use altogether. Hardware needs to be carefully specified, designed and monitored to ensure acceptable performance.

References

1. Anderson M. The toll of downtime: a study calculates the time and money lost when automated systems go down. Healthc Inf. 2002;19:27–30.
2. Laprie J. From dependability to resilience. [Online]; 2008 [cited Jul 2015]. Available from: http://2008.dsn.org/fastabs/dsn08fastabs_laprie.pdf.
3. International Electrotechnical Commission. IEC/TR 80002-1:2009. Medical device software, Part 1, Guidance on the application of ISO 14971 to medical device software. 2009.
4. Magrabi F, Ong M, Runciman W, Coiera E. An analysis of computer-related patient safety incidents to inform the development of a classification. J Am Med Inform Assoc. 2010;17(6):663–70.

Part III
Getting Your Organisation Ready

Chapter 8
The Safety Management System

8.1 What Is a Safety Management System?

Early approaches to safety management relied on incident reporting and the implementation of corrective actions to prevent recurrence [1]. Whilst this is a first step in managing safety, a reactive method has a number of shortcomings. Firstly one has to wait for an incident to occur and suffer the resulting harm before the cause can be addressed – hardly a strategic approach. Secondly in industries with low failure rates such as aviation it becomes increasingly difficult to reduce risk by this approach alone. Essentially the safer something becomes the less there is to measure and therefore learn from.

A proactive strategy is the principle alternative to a reactive method. This line of attack employs prediction taking into account what we know about the system at any point in time, its users and the environment into which it is deployed. Risk reduction activities occur not just in response to adverse incidents but strategically during the product's design, manufacture and operation. This is achieved through the introduction of a Safety Management System (SMS) into the product's development lifecycle. The SMS comprises the processes, activities, governance and resources to identify, assess, evaluate and mitigate the clinical risk associated with an HIT product. In other words the actions needed to develop and maintain the safety case.

Formulation and maintenance of the safety case requires co-operation, communication, logic and planning. These are all characteristics of a well thought-out, considered and documented SMS. Clinical Risk Management (CRM) fundamentally cannot be undertaken in an ad-hoc and piecemeal manner. To do so would inevitably result in an incomplete, unjustified or invalid safety case. But the presence of a process brings more than just a way to organise one's thoughts. Being seen to follow process is as important as following the process itself when it comes to critically evaluating its performance.

A. Stavert-Dobson, *Health Information Systems: Managing Clinical Risk*,
Health Informatics, DOI 10.1007/978-3-319-26612-1_8

The application of CRM drives the mitigation of clinical risk. Like all controls, this mitigation needs to be evidenced and the presence of a formally documented SMS can offer that assurance. To approach a CRM assessment in an ad-hoc manner invites criticism of the methodology used. This can result in unwelcome and unnecessary noise in what might otherwise be a sound analysis. By establishing, agreeing and documenting the methodology up-front stakeholders have an opportunity to influence and buy in to the approach even before a particular HIT product is considered.

The development of an SMS also brings the benefits of repeatability and predictability. The majority of CRM assessments will need to be revisited and repeated from time to time. This might occur when an organisation develops or encounters a new product or version or when that same product is rolled out to new care settings. To formulate and document the process that will be followed each and every time one goes through the loop involves unnecessary administrative overhead and mental effort. By developing a single, agreed SMS one can retrace the required steps safe in the knowledge that the methodology is sound and by that point proven.

Repeatability brings additional benefits when the SMS is audited. The auditor is able to drawn conclusions by sampling material from a random project and extrapolating the findings to other projects where the same methodology has been adopted. An ad-hoc approach requires a deep-dive into each project typically with different findings. Where a non-conformance is found this can often be managed by a simple change to the underlying process with the comfort that, so long as the new process is followed, all subsequent projects will conform.

Organisations may decide (or be required) to operate an SMS which is compliant with a specific local or international standard. By developing a consistent process that is compliant with the standard or standards, its requirements can be flowed down and inherited by the SMS. Those who are undertaking a CRM assessment can follow the SMS without necessarily concerning themselves with the specific details of individual standards – ultimately achieving compliance by abstraction.

The implementation on an SMS requires the input of a number of stakeholders. Clinical experts, software designers, testers and project managers will all contribute to material that will form the basis of the CRM analysis. The timely delivery of that material is a critical dependency on the ability to produce a valid and complete safety case. As with any project, these 'gives and gets' need to be plumbed in to the high level project plan and managed accordingly. By implementing an agreed process, this simplifies the identification of these touch-points. Frankly, individuals simply get used to working with the SMS, know what information they need and what expectations there are. This comforting situation simply cannot develop where an ad-hoc, free-for-all approach to CRM is taken.

A CRM assessment will generally be a cost to the project (although of course one can argue the ultimate value add of the assessment more than recuperates that cost). The cost needs to be included in the project business case but quantifying it early in the product lifecycle can be a challenge. By developing a repeatable SMS, one is able to better predict the likely effort involved in applying that process to a particular product.

Finally note that an SMS is not the same as a safety case. The SMS sets out the processes and methodologies that an organisation will harness in pursuit of building a safety case. In other words the safety case is one of the outputs of applying the SMS to a particular HIT solution or component. Whilst the SMS will define acceptability criteria it will say nothing about whether the risk profile for a specific system is tolerable.

8.2 Types of Safety Management Systems

SMSs typically fall into two categories; prescriptive and goal-based systems.

The prescriptive system is the simplest to understand and centres on inspection. With this methodology, one defines, adopts or inherits a series of binary criteria or requirements often set out as a checklist. The criteria are applied to the target product or service and, if all requirements are met, the product is deemed 'safe'. If any requirement is not met steps are taken to address the specific problem and once dealt with the box is ticked. The criteria may be broad or specific and, in some cases, may be prescribed by a regulatory authority, commercial contract or procurement rule.

This approach is simple – or maybe simplistic. Whilst the developers of the safety criteria need to be experts, those applying the criteria need not be. An entire assessment can sometimes be undertaken in a matter of minutes. As long as the information is available to make an assessment against the criteria, one need not question the rationality or validity of the process itself. The clear danger is that the risk management becomes nothing more than a 'tick-box' exercise with little regard for the big picture.

The validity of the prescriptive approach depends on the ability to generate criteria which are appropriately sensitive and specific. This is where the methodology quickly becomes unstuck. The first assumption is that a positive response to a requirement will translate into a genuinely safe system characteristic. For example, one might propose that the use of HL7 messaging as the basis for cross-system communication represents an industry standard and state of the art engineering. The use of this technology might therefore seem a reasonable criterion to contribute to a system being deemed 'safe'. In reality of course, the use of HL7 says little, if anything, about the safe operation of a system – it simply states conformance with a particular technical standard. It would be perfectly possible to generate a valid HL7 message which is subsequently processed in an unsafe way.

A similar but opposite assumption is also in play with the prescriptive approach. The presence of a criterion may imply that the characteristic in question is the only means by which the system can be rendered 'safe'. Continuing the HL7 example above, there is the assumption that the use of any other communication standard would be intrinsically associated with intolerable risk. Clearly this logic does not follow. A system may use a perfectly reliable vendor-specific message format, one that the manufacturer has utilised for many years and has a great deal of operational experience with. Indeed a change in technology to meet the requirements of the prescriptive system may in itself introduce unnecessary risk. The prescriptive system can pointlessly stifle innovation, and safe innovation at that. Design priorities become about meeting arbitrary targets rather than strategically addressing specific hazards. Any cost and time-saving benefit offered by the approach are soon lost when one considers the effort needed to maintain and validate the underlying criteria.

Having said all that, some success has been achieved in formulating suitable checklists as a means of self-assessing the safety of Electronic Health Records (EHRs). Of note are the US SAFER (Safety Assurance Factors for EHR Resilience) Guides issued in 2014. These documents cover nine distinct aspects of EHRs and recommend a set of system characteristics intended to optimise safe implementation and use of the technology [2]. Further research is needed to determine the effectiveness of this material and the approach proposed.

In contrast to the prescriptive method is the goal-based approach which has very different characteristics. With this methodology an organisation decides on the ultimate aim of undertaking a CRM assessment and defines the necessary steps to assess the achievement of that goal. The goal is defined in a CRM Policy and specifies the degree of clinical risk which is acceptable. A process defined in the SMS is then applied which teases out the hazards and controls for a product and establishes the clinical risk profile for the target HIT system. The outcome of the analysis is then tested against the goal, i.e. whether or not the established risk is acceptable. The argument and evidence is often made available to appropriate stakeholders to form their own conclusions on the claims made.

Note that a key difference between the approaches is that in a goal-based analysis the responsibility for determining the assessment methodology, risk acceptability and justification remains with the developers and operators of the SMS rather than with a regulator or other external body. An external party might specify a need for a goal-based SMS and visibility of the output but not its content or ultimate objective.

The technique requires a methodical and thorough approach taking into account all relevant information from the design, build, test and live operation of the system. In this way, the system is considered on its merits given its intended purpose and design. A key feature of this approach is the gathering of appropriate evidence to substantiate the risk assessment. The safety argument, supporting evidence and conclusions are then set out in a safety case.

Unlike the prescriptive approach, the goal-based methodology enables an assessment and re-assessment of risk as operational experience is gained. The intelligence gathered after system go-live provides powerful lessons from which the safety position can be re-evaluated and further evidenced. A simple prescriptive checklist fails to take these important messages into account.

Note that the approach does not preclude compliance with any specific regulatory or standard requirements. Rather the SMS should ensure that compliance is achieved in the course of safety case development. Essentially, the safety case represents an information superset and knowledge base potentially facilitating compliance with any number of relevant safety standards.

The disadvantage of the goal-based approach is that it can be more complex to apply and requires personnel who are skilled in the technique. Compared to a prescriptive approach, a deeper understanding of the system's design and operational domain is usually needed in order to present a valid and justified argument. All of this adds to the cost, effort and time involved in performing the assessment. The benefits, are though, substantial and there is a tendency to migrate to goal-based approaches in many areas of safety regulation [3]. The rest of this book will focus on the development and application of goal-based SMSs.

8.3 Documenting the Safety Management System

Principally there are at least couple of ways of documenting the SMS. Either the approach can be set out in one or more artefacts to which the individual CRM project documents refer or the process can be explained in the CRM project deliverables themselves. Each method has its merits depending on the nature of the project.

Where an organisation has a single product which is regularly subject to CRM activities, it is not unreasonable to discuss the CRM methodology exclusively in the safety case itself. In this way, each section of the report essentially reads as:

- We needed to do this…
- So here is what we did…
- These were our findings…
- And the conclusions that can be drawn are…

The advantage here is that the entire approach can be understood (and therefore evaluated) in a single document forming a logical and complete argument. Those who consume the safety case may well be grateful for the succinct and comprehensive approach. However it is important to understand this methodology does have some shortcomings.

The situation can quickly become complex when one has to evaluate a number of products or several modules of a single product. When an organisation generates multiple safety cases on a regular basis, having to discuss the methodology each time can become cumbersome. If the process varies between projects, this can be difficult to appreciate without a side-by-side comparison. If the same process is described in different ways across multiple documents this can lead to confusion and difficulties in version control. Occasionally one has a need to conduct a very brief CRM assessment, perhaps a small change to a product for example. Having to set out the process in each safety case update can create far more content than the actual assessment results.

Finally, the development of a rigorous SMS can be a valuable asset to an organisation. Wrapping this intellectual property inside a safety case (which can have a wide distribution) may disclose this asset to a wider audience that would otherwise be desirable.

The adopted approach should therefore be carefully thought through taking these issues into account. In the main, if an organisation intends to implement any more than a couple of brief safety cases, it is probably worth investing in creating a set of dedicated documents which describe the CRM process from start to finish and have separate documentation which sets out the result of applying the process to each HIT product.

In terms of SMS content, the documentation should at a minimum contain the following:

- The CRM Policy, terms of reference and definitions
- Process overview
- Risk acceptability criteria
- The structure and governance of the CRM team, its position in the organisation, accountability and competency assessment methodology
- List of deliverables, their purpose and integration into the product lifecycle
- An account of the detailed processes required to identify, assess, evaluate and validate hazards.
- Tools and equipment needed to apply the processes
- The document management and sign-off strategy

The SMS documentation should exist in a single, easy to access location – typically this will be managed electronically. From time to time it will become necessary to update the SMS documentation and this should be managed through a formal process of version control, review and sign-off with clear traceability as to what was changed, when and why.

Note that the SMS documentation will subsequently form the basis of an SMS audit. The documentation should therefore be authored with this in mind to ensure that each requirement and claim made by the system documentation can be easily evidenced and evaluated.

8.4 The Clinical Risk Management Policy

An SMS is more likely to be credible in organisations with an explicit recognition of patient safety and where CRM is formally expressed as objectives. A risk management policy expresses an organisation's commitment to risk management and clarifies its general direction or intention [4]. The development of a clear CRM policy, agreed and signed off at the very top of an organisation sends an unambiguous message to its employees and external stakeholders. It makes clear that CRM is the responsibility of everyone irrespective of their role and provides practitioners with a mandate to act when safety is under threat. In this way, the policy contributes to an organisation's strategic vision, values and stability.

To begin with, organisations need to decide at a most fundamental level whether or not they intend to undertake CRM activities at all. After all, not all HIT software is capable of impacting patient care (as discussed in Sect. 1.6.1). Even in these circumstances though it may be appropriate for an organisation to develop a policy setting out this position and the underlying rationale. Similarly, whilst an organisation may not benefit from CRM activities currently should their product portfolio or intended purpose change (intentionally or otherwise) CRM may be required in the future. The policy in this case should define how this situation will be re-assessed, when and by whom.

Where CRM is required within an organisation a policy should be developed to reflect this requirement. Without such a policy, those responsible for driving CRM activities lack a mandate to take action and such a situation results in an absence of authority, credibility and the capability to influence. Occasionally CRM stakeholders can suffer from lack of support, difficulty in communicating their goals and without a clear escalation route or sense of direction. An organisational-wide CRM policy provides individuals with something to point at, gives reason and rationale to their actions and provides a firm basis for objective setting.

Assuring a product is not limited to those individuals in the CRM team. It will be necessary to gain buy-in from clinical representatives, system testers, project managers, training staff, etc. Typically those individuals will not reside in the direct CRM chain of command. A dynamic and energetic CRM team is always a bonus but that enthusiasm may not be shared by those who have other objectives, responsibilities and targets to meet. It is not always clear to stakeholders why CRM is important in the big picture and of the clinical and business benefits it provides. The presence of a policy brings those diverse stakeholders together and gives them collective responsibility to cooperate in achieving a common goal.

The safety assurance of a system inevitably requires resources and in some cases the demand is not insignificant. Failure to develop a CRM policy can deprioritise these activities at times when resources are challenged. All too quickly one can end up in a situation where CRM effort is completely omitted from a business case entirely leading to project risk, unexpected costs and potentially patient harm.

At a minimum, the policy should contain:

- The organisation's objectives and rationale for conducting CRM activities along with their commitment to manage safety effectively. Ideally this should dovetail into an organisation's wider values and purpose.
- An indication of how safety will underpin the work of the organisation and the means by which this will be achieved.
- Reference to the standard, regulatory framework or contractual requirements with which the SMS will comply.
- A summary of the methodologies that will be harnessed to undertake each assessment and references to the documentation which will detail the processes employed.
- A commitment to provide resources to support CRM activities and an indication of the team who will be responsible for implementing the system, their terms of reference, position in the organisation, remit, etc.
- A summary of how the CRM system will be governed, what the chain of command will look like and what forums will exist for top management to monitor performance of the SMS.

- A clear indication of who the key decision makers are, who has ultimate accountability for safety and for ruling on the recommendations of the CRM team.
- How individuals within an organisation are expected to prioritise safety in their day to day work as part of a safety culture and how concerns can be raised in a blame-free, non-prejudicial manner.

8.5 The Safety Management System Processes

Organisations can find it difficult to get started in implementing an SMS. The key is to begin with a solid, well thought-out and accurately documented process. The process should align to the complete end-to-end lifecycle of the manufacturing and/or deployment project with key touch-points at project milestones. One should be mindful that a rigorous but disconnected methodology can struggle to gain credibility, visibility and acceptance in an organisation. Process developers should therefore identify existing improvement processes and programmes within their organisations and harness established skills, expertise and good-practice. By minimising change, processes which are extensions of current operation are likely to be easier and more cost effective to implement.

The CRM process should be implemented as early as possible in a project as this will allow key design decision to take potential controls into account at the outset rather than suffer unexpected rework costs later. Basically, the sooner CRM activities become business as usual, the more effective and cost-efficient the work will be. Developing this process should involve the consultation of all appropriate stakeholders ensuring that the various 'gives and gets' align and avoid circular dependencies. The use of a RACI matrix (Responsible, Accountable, Consulted, Informed) [5] or similar tool can be helpful in deriving these relationships.

The detailed processes will describe the interactions and interfaces between stakeholders in order to gather, evaluate and document the safety analysis. Organisations should give thought to the appropriate level of granularity for these documents depending on the complexity and risk associated with the systems assured. Ultimately the material should represent a complete and detailed set of instructions and workflow to undertake a CRM analysis and create the safety case. Importantly, the SMS processes should identify clearly who does what and when – leaving no doubt as to where individuals' responsibilities and expectations lay. Whilst this may be a significant effort overhead initially the investment is likely to quickly pay off when one considers the time involved in training new staff and driving a consistent approach across an organisation.

Whilst the nature of the SMS process documentation will differ between organisations, in a typical HIT SMS one might expect to see instructive material in the following areas:

- Developing the CRM plan
- Developing a hazard register

- Managing safety related requirements
- Developing a safety case
- Managing system change
- Safety testing and clinical evaluation
- Monitoring the system in live operation
- Assessing personnel competency

At a minimum a process should include:

- A clear objective for the process and when it is carried out.
- Background to enable a reader to put the process into context and identify when it may be appropriate to modify the approach to implementing the process.
- Any pre-conditions that should exist before the process can be executed.
- A step-by-step explanation of what should be done in clear, unambiguous language.
- The actors involved in performing each task and what the expected outcomes are.
- The resources and tools which should be used, where they can be found and what training material exists for them.
- A conclusion setting out what should have been achieved at the end of the process.

8.6 Governing the Safety Management System

8.6.1 Establishing Accountability

Without a clear chain of command ultimate responsibility for clinical safety can be blurred. Dealing with a legal challenge is not the time to begin ascertaining accountability. Any organisation developing an SMS needs to ask, who is ultimately accountable for safety? Who decides whether a product can go live? Who decides whether or not a costly, safety-related defect is fixed?

Whilst the CRM team is responsible for evaluating and characterising risk often these individuals lack the authority to have ultimate decision-making ability. The CRM team may ascertain that the risk associated with a product go-live is unacceptably high but their conclusion is likely to be no more than a recommendation. The final decision almost certainly sits with senior management who also have legal, commercial and reputational responsibilities. Whilst they would do well to heed the advice of an experienced and objective CRM team, it is senior management who are accountable for an organisation's actions and not (at least exclusively) those who were appointed to evaluate the risk. Perhaps the exception is where those undertaking the assessment have professional and legal responsibilities (such as chartered engineers). Whether this extends to clinicians and their professional registration is currently unclear.

So who is best placed to be accountable for CRM? Realistically this should be a senior manager, typically the most senior member of an organisation who is able to have an understanding of the domain. This could be the chief executive, chief operating officer or medical director. Their role is typically one of governance and of overseeing the safety activities. For example;

- Establishing a CRM Policy
- Monitoring the effectiveness of the Policy
- Determining the organisation's acceptability of risk
- Overseeing the development and implementation of the CRM system
- Final approval of deliverables
- Provision of appropriate resources

One particular challenge for personnel governing a CRM system is the involvement of third-party organisations tasked wholly or partly with the responsibility for undertaking CRM work. In most cases organisations will not be able to transfer their risks and responsibilities onto third-parties and therefore an environment of close co-operation and managed expectations is warranted. In particular, where a manufacturer is required to comply with particular regulatory standards, it would be wise to flow down appropriate performance requirements to any subcontractors.

8.6.2 The Safety Management Committee

A clear chain of command and escalation route are essential to operating an efficient SMS. It is however also beneficial to augment this with some kind of regular forum which acts as a single point of reference to facilitate system governance. Whilst top management take ultimate accountability for safety, a collaborative approach is able to pool ideas and experience, manage issues and assist in the forming of objective, evidence based judgements.

It is for individual organisations to define the makeup of the committee but typically it might include product experts, CRM leads, clinicians, project/programme managers, etc. along with the accountable senior manager. Some organisations choose to include patient representatives to offer a degree of independence. The forum can be internal to an organisation or involve external stakeholders such as customers or suppliers. The terms of reference of the group should be clearly defined and documented – this might include the committee's purpose, tasks, responsibilities and membership. It may or may not be appropriate to cover the day to day project management business at the session depending on the size of the organisation, representation, etc. Certainly a focus of the session will be to examine those risks and issues which are of concern and require senior management escalation.

Part of the committee's responsibility should be to monitor the effectiveness of the SMS (see Sect. 8.7.2) and identify any shortcomings or inefficiencies. Changes to the system should be managed through the committee to identify potential impacts. The meeting should be accurately minuted (preferably by a dedicated minute-taker) and the minutes distributed after the session for agreement. This document should then be archived and be available to be called upon in any audit which may subsequently be undertaken.

8.7 Auditing and Evaluating the Safety Management System

The effort which goes into creating an SMS is an investment for the future. But to ensure a return is received on that investment it should be accompanied by a solid commitment to evaluating its effectiveness and showing continual improvement over time. Evaluation is about determining whether the system achieves its objectives and this analysis can reveal enlightening gaps which need to be addressed for the SMS to perform well. The evaluation should:

- Be a recurring and planned activity
- Involve suitably experienced personnel
- Have a clear scope
- Be time-bound
- Have agreed deliverables and objectives.

There are two key approaches to evaluating the system and both are necessary to obtain a rounded view. The first is a simple audit of the system itself. This can be undertaken at any stage of establishing an SMS but it often pays particular dividends when a system is still maturing. Secondly one can investigate the effectiveness of the SMS in the real-world, in other words to determine whether the SMS is successful in managing risk introduced by the HIT system.

8.7.1 Auditing the SMS

As a starting point for the audit one needs to decide exactly what the SMS is being audited against. Where the intention is to comply with a particular standard, this can be a straightforward task of testing the requirements against the SMS and its implementation. Where there is no such standard to comply with, there is value in investing time in creating quality criteria which must be met in order to fulfil the requirements of an organisation's safety policy. Without criteria being in place it is difficult or impossible to have an objective measure and therefore to be able to pinpoint where gaps in the SMS exist.

The individual who carries out the audit might be external to the organisation being audited (especially where formal certification is being sought) or be a regular member of staff. Carrying out an internal audit is a valuable learning exercise and is relatively easy to execute. An audit does however require some particular skills and where internal personnel are used it is important to ensure that they are competent to undertake the task. The auditor will need to be familiar with the criteria being audited against and possibly have some experience of the HIT domain. However, an individual who is also involved in the project operationally or who has interests which might conflict with the outcome should perhaps be substituted for someone who can be seen to be more objective.

Audits can take many forms but generally they will involve one or more interviews between the auditor and key personnel. The subject is posed a series of questions about how they work, the procedures they follow, the training provided and level of supervision. The auditor will expect the subject to be able to show evidence of the claims they make, examples of their work, plans, meeting minutes, project materials and organisation charts. Most of these materials will need to be easily accessible at a moment's notice to reflect the fact that they could be called upon at any time operationally.

An audit can essentially be split into two areas; an evaluation of the SMS's composition and an appraisal of the SMS's implementation.

An audit of the SMS composition will evaluate its structure and content to determine whether the necessary processes and procedures to manage risk have been thought through, agreed and documented. For example, an auditor might ask:

- Is there a clear safety policy in place, signed off by senior management?
- It is clear who is accountable for safety and who is responsible for undertaking a risk analysis?
- Are the SMS processes fully agreed, documented and signed-off?
- It is clear how risk will be estimated and evaluated?
- Has it been ascertained what level of risk is acceptable?
- Is the SMS readily accessible to those who need to implement it?
- Have resources been allocated appropriately?
- Do individuals who take responsibility for implementing the SMS have the right skills and training?
- Is development, revision and maintenance of the SMS captured in a time-bound project plan?

The focus here is on ensuring that those who implement the system know what is expected and can readily ascertain the means by which the SMS's objectives are achieved. Essentially it's an evaluation of the instruction manual but stopping short of how it is put into use in the real world.

In contrast, when an auditor inspects the implementation of the SMS the assessor looks at its application to one or more HIT products, the deliverables generated and the completeness of the safety projects.

For example, for each HIT system being assessed an auditor might ask themselves:

- Is there a CRM plan in place, is the plan agreed, signed-off and being kept up to date?
- Have each of the processes described in the SMS been applied and followed during the project? Did the application of those processes conclude?
- Have appropriate deliverables been authored and are future deliverables suitably planned?
- Are the deliverables of a good quality, are the documents version controlled, stored in the correct place and approved?
- Is it clear which personnel are undertaking the CRM work, has their competency been assessed and evidenced?

- Have regular project meetings taken place, are the minutes available, do the minutes contain time-bound actions with clear owners?
- Does application of the SMS drive out clear conclusions, and recommendations?

In practice, the auditor may choose to inspect both the structure of the SMS and its implementation at the same time. For example, he/she might look at the process for creating a hazard register and note a requirement for its delivery to be aligned to a project milestone. Compliance with this might be demonstrated by showing an example of a project plan for a particular assessment and evidence of delivery to that milestone. The hazard register in question may then be called up, its issue date and version number checked and the approvers validated. In this way the auditor is able to trace from the quality criteria to the process to its implementation in a simple and logical way.

At the end of the activity the auditor will document his/her observations and identify any non-conformances. The project team must then plan to address the non-conformances with appropriate corrective actions. Again this must be a planned and time-bound activity with clear goals. The auditor may wish to see evidence of those corrective actions and provide comment on the final status of any non-conformances.

The process of audit should not be seen as a punitive or box-ticking activity but rather a rich and valuable learning exercise to validate the work to date and find opportunities for continuing development. Experiencing an audit is also likely to change the way in which future documentation is written and evidence preserved. A well-documented SMS will lend itself to being audited not only making the job of the auditor easier but also to facilitate preparation and build confidence of those individuals being interviewed in the process.

8.7.2 Evaluating Effectiveness

Effectiveness, in contrast to simple audit, looks at the outcome of implementing the SMS. In other words, does the SMS actually succeed in minimising the risk associated with creating and implementing a HIT system? This can be a challenge to prove objectively, not because SMSs are typically ineffective but because the process of capturing relevant data and finding useful metrics can be troublesome.

One of the key challenges is in establishing the baseline one intends to measure success against. Ideally one would want to implement a HIT system in two identical environments, one with and one without the application of an SMS. In practice this is impractical on many different levels. Of course where an SMS is introduced alongside an existing and establish HIT system one has an opportunity to observe any changes on reported adverse incidents. However one must bear in mind that in this case the opportunities for practical risk reduction in a mature system are relatively limited and demonstrating a statistically significant improvement in safety could require a substantial amount of effort.

There are a number of metrics that one could harness as a proxy of SMS effectiveness depending on the nature of the product and care setting. For example, the longitudinal number of prescribing errors might be interesting to monitor in an electronic prescribing system. Often the most valuable learnings can be obtained by observing not an absolute measure at a single point in time but rather a trend over a longer period. Many metrics suffer from being contaminated by extraneous factors outside of SMS performance but by observing changes over time one is able to ascertain at a basic level whether things are improving or deteriorating.

The ultimate goal of the SMS is to minimise the occurrence of harm to patients where the HIT system is the cause or a contributor to the events leading to harm. Most healthcare organisations now implement adverse incident reporting processes (see Sect. 5.5) so an opportunity exists to analyse the numbers and types of incidents over time. Differentiating those events where technology played a role can be challenging to ascertain where incident management systems fail to capture this data specifically. Nevertheless the advantage here is that the process of gathering SMS performance data harnesses existing and mature reporting methods.

Using incident reporting to evaluate SMS effectiveness has limitations when it comes to identifying low level adverse events. For example, where despite a hazard being triggered, harm was prevented through human intervention or where a problem occurs so frequently that completing an incident report each time simply isn't practical. In these cases, observing reported adverse incidents alone may fail to reflect the true impact of the SMS.

One alternative metric is the number of safety-related system issues which are reported to the support service desk. In this way, instead of looking at the adverse events which have a technological causative component one inverts this and considers which faults could adversely impact care. A successful SMS should result in a product and implementation which minimises safety-related faults so this may be a reasonable analogue of the SMS's performance. Most healthcare organisation operating HIT systems offer some form of user support and facilities for fault logging. There are many benefits in evaluating which faults could compromise safety (see Chap. 19). This method has advantages in that it is not necessary for patient harm to have occurred for the issue to be detected. The technique can be labour intensive however as each reported fault must be assessed for potential safety implications and those reporting it may not have sufficient knowledge of the erroneous behaviour and its scale to make this judgement.

A simple but enlightening technique for estimating SMS effectiveness is the employment of a staff survey, especially where this is used to supplement other data. The advantage of a survey is that it provides an opportunity to gather information at a granular level. For example, one could ask users to score questions such as:

- How frequently do you observe circumstances where the system contributes to adverse clinical incidents?
- How many on-going issues are you aware of which have the potential to impact care?

- When you detect a fault which has the potential to adversely affect care, how easy is it to report the issue?
- Did the training you receive on the system help you to understand where there might be safety implications?
- Do you have access to materials which help you to understand how the system should be operated?
- When changes have been made to the system, did you receive sufficient notice and training?
- Do you have opportunities to contribute to the future safe design and configuration of the system?

Surveys of this kind are of course subject to bias. Individuals are perhaps more likely to respond if they unhappy with the system generally or the way that it has been implemented. As with the other metrics, trends observed over time are more informative than taking a snapshot view.

One final metric that may be chosen to assess SMS performance is the level of residual risk itself. It could be argued that an optimal SMS will ensure that all hazards are mitigated to as low as reasonably practicable thus where there are risks which are insufficiently controlled this represents deficiencies in the process itself. This approach can be attractive to manufacturers who don't necessarily have access to the outcome metrics available to healthcare organisations.

One should however apply some judgement in the precise measure being taken to correlate with uncontrolled risk. For example the absolute number of hazards identified may simply reflect the detail and rigour of the assessment process rather than any underlying residual issues. Similarly the absolute number of test defects logged may be a result of diligence rather than being a comment on product quality. One could however reasonably decide to measure the number of unfixed safety-related defects, above a certain level of risk, at a key project milestone. For many programmes this will be a reasonable analogue for residual risk and SMS effectiveness. Alternative measures available to manufacturers might include the number of unresolved safety-related service desk calls or the number of on-going project issues which could impact care.

8.8 Managing Process Variability

In the ideal world one would be able to develop a single process that could be efficiently applied to every HIT product which an organisation manufactures or implements. In reality, things are not quite so straight-forward; there are occasions where it is necessary to vary the approach somewhat. This does not mean to say that one automatically has the right to recklessly break away from an established process into which much thought and effort has been applied. Rather, a good process in itself should contain a degree of flexibility – a process which can be moulded and tailored to the needs of a specific project.

The need to vary a process from one project to another can occur for a number of reasons. These include:

- The stage the target product has reached in its lifecycle
- The scope and complexity of the project
- Whether the product is being developed from scratch or is Commercial Off The Shelf (COTS).
- The project timescales
- The presence or absence of previous assessments
- The nature of the implementation or operational environment

In addition there is at least one more common and justified reason for adapting the process – a variation to better reflect the degree of rigour needed to assure the product. The Collins English Dictionary includes in its definition of rigour "Strictness in judgment or conduct" [6]. Essentially the term refers to the amount of effort one is required to invest in safety assuring a system given the anticipated degree of risk.

This balance between risk and rigour is a key principle in all SMSs. The right balance is ultimately a judgement call but a mismatch either way can have undesirable results. At times one can be tempted to continue to mitigate risk just because one can; one more policy, one more training module, a further round of design assurance, just a little more testing. These activities may be appropriate but it is prudent to keep in mind the justification for their undertaking. In each case, that justification should be founded on the established or expected degree of risk associated with the system in question. Failure to adopt this approach can lead to significant effort for little gain, for want of a better word, waste. Conversely in the midst of product development or implementation one can lose sight of those systems or areas which would genuinely benefit from additional effort and rigour.

When a variation to the process is considered it is worth separating out in one's mind whether the driver is a practical measure to meet the requirements of the project or to better reflect the degree of rigour required. For example, suppose the timescales associated with a project are tight and the anticipated level of risk associated with a system is low; one might justify a decision to omit one or more CRM activities. On the other hand, omitting process steps on ground of timescales alone without any consideration of the impact on rigour would potentially be foolish as this could lead to an incomplete, unjustified or invalid safety case at a late stage in the project.

One could argue that until the assessment has been undertaken, how can the level of risk be ascertained? Without knowing the level of risk, the required degree of rigour cannot be evaluated and therefore how can variations to the process be established at the start of the project? This is always a difficult conundrum but it re-enforces the need to develop a high-level view of the potential clinical risk early in the project. In many circumstances this can be estimated based on knowledge of the care setting and the clinical business processes the system intends to support. For example, the clinical risk associated with an electronic prescribing system is likely to be greater than that associated with an equipment tracking system. Particularly where an organisation manufactures or implements a number of different systems, this approach can help to focus effort and resources and appropriately match risk with rigour.

Where a variation to the process is considered, this should be formally documented in the CRM Plan for each system under assessment along with the rationale and justification for the variation. The process documentation itself could also set out any basis for omitting relevant steps. Building in process flexibility at the time of SMS development facilitates the straight-forward justification of the approach when one applies the process to a product.

8.9 Document Management

Large CRM projects can generate a significant amount of documentation. These valuable assets need to be managed with great care and measures put in place to ensure their availability and backup. Some Standards such as ISO 14971 [7] and ISB 0129 /0160 [8, 9] specifically require that a risk management file be formally maintained. It is this file which contains the project deliverables and supporting evidence which, taken collectively, represents the body on knowledge to justify the product's safety profile.

In practice, the file needn't be a physical paper folder or filing cabinet but more usually an electronic repository or document management tool. Maintaining this collateral in a single (virtual) place greatly facilitates audit and supports an organised approach. Also note that in many cases much of the content of the file will be material which is not specific to CRM but pertains to the project more generally. It is common practice to refer and reference out to this material rather than provide duplicated copies within the file itself (and risk the loss of version control).

In large organisations it is common for a general-purpose document management system to be implemented – in most cases it makes sense for CRM materials to harness these tools. A document management system will typically set out the strategy for:

- Document registration and coding
- Version and configuration control
- Storage, retrieval and backup
- Editorial standards
- Common terms and definitions
- Document templates, branding and copyright notices
- Peer review and sign-off processes

8.10 Clinical Risk Management Training

The need for CRM training, a proposed methodology for its delivery and collateral which supports it form an integral part of the SMS. Demonstrating that CRM processes are underpinned by effective communication and socialisation adds significant credibility to the approach and raises confidence in its rigour come an audit. Consideration should be given as to:

- Who delivers the training and to what extent is their competency assessed
- What form will the training take, for example face the face sessions, eLearning, Webinar, etc.
- How those who need to be trained will be identified
- How the variety of CRM training needs across an organisation will be assessed and delivered to
- How completion and effectiveness of training will be assessed
- How attendance at training will be recorded and absentees managed
- How new starters will be briefed
- How frequently training materials will need updating and refresher sessions executed

8.11 Summary

- An SMS underpins an organisation's approach to managing risk in its HIT systems.
- Goal-based methods are preferable to prescriptive approaches. They allow risk to be objectively justified and explained in the form of a safety case.
- The SMS will be associated with a CRM Policy and a set of processes which outline how risk assessment and reduction will be achieved.
- The policy should set out how the SMS will be governed and who is ultimately accountable for the conclusions drawn.
- The integrity and effectiveness of the SMS needs to be evaluated from time to time and any shortcoming dealt with.
- The approach to CRM will vary from project to project and the SMS should be flexible to support this. The primary driver for variation should be the estimated degree of risk associated with the system under examination.
- The SMS and its application can generate a significant amount of documentation in large projects. This needs to be carefully administered to maintain quality and availability.

References

1. Bayuk A. Aviation safety management systems as a template for aligning safety with business strategy in other industries. Creative Ventures International, LLC., 400 South 2nd Street, Suite 402-B, Philadelphia, PA 19147. http://www.scribd.com/doc/221557284/Safety-Mgmt#scribd.
2. The Office of the National Coordinator for Health Information Technology. SAFER guides. [Online]. 2015 [Cited Aug 2015]. Available from: HYPERLINK. http://www.healthit.gov/safer/safer-guides, http://www.healthit.gov/safer/safer-guides.
3. Penny J, Eaton A, Bishop P, Bloomfield R. The practicalities of goal-based safety. Proceedings of the Ninth Safety-Critical Systems Symposium Bristol, UK, 6–8 Feb 2001. Springer, 2001 ISBN: 1-85233-411-8; 2001. p. 35–48.

4. International Organization for Standardization. ISO 31000:2009, Risk management – Principles and guidelines 2009.
5. Smith M, Erwin J. Role and Responsibility Charting (RACI). [Online]. [Cited Jun 2015]. Available from: HYPERLINK. http://pmicie.org/images/downloads/raci_r_web3_1.pdf, http://pmicie.org/images/downloads/raci_r_web3_1.pdf.
6. Collins Dictionaries. [Online]. Available from: HYPERLINK. http://www.collinsdictionary.com, http://www.collinsdictionary.com.
7. International Organization for Standardization. EN ISO 14971:2012. Medical devices. Application of risk management to medical devices. 2012.
8. Health and Social Care Information Centre. ISB0129. Clinical risk management: its application in the manufacture of health IT aystems version 2. UK Information Standard Board for Health and Social Care. London. 2013.
9. Health and Social Care Information Centre. ISB0160. Clinical risk management: its application in the deployment and use of health IT. UK Information Standard Board for Health and Social Care. 2013.

Chapter 9
The Scope of Clinical Risk Management

Once financial, environmental and reputational forms of risk are excluded one might expect that what is left is a fairly tight definition of the factors which constitute clinical risk. Experience has shown however that there are many boundary cases that can be subject to debate. Clinical risk arises from the manifestation of one or more hazards. Hazards have an intrinsic relationship with harm but it is often in the definition of what constitutes harm where debate arises.

In any CRM assessment one needs to define the boundaries of the analysis (as discussed in Sect. 11.1.1). Those systems, modules, components or areas of functionality which have the potential to impact care are often described as being 'safety-related'. Thus, an entity can be considered safety-related if it:

- Influences the delivery or nature of clinical care, and
- Affects care in such a way that (through action or inaction) it could contribute to harm.

Note that an important word in this criterion is 'contribute'. The system does not need to be directly responsible for harm to make it safety-related. In the majority of cases there will be a human being between the system and the patient. This individual will typically have professional responsibilities and be required to apply clinical judgement. Essentially, an HIT system can only be one contributor to a chain of events that result in harm.

Many examples of potential harm are clear-cut; a patient is given the wrong drug, fails to be invited to a crucial appointment or unnecessarily requires re-investigation. However, there are a number of boundary cases where the definition of harm is blurred. Whether or not these boundary cases constitute harm is, in many ways, a philosophical, ethical or moral argument and it is this which can lead to difficult conversations. Perhaps a more practical approach is to determine whether managing boundary cases can best be achieved through CRM activities or by some other means. This quickly circumvents the problem where we recognise that a goal is important but its links to safety are blurred. What is key for any organisation

A. Stavert-Dobson, *Health Information Systems: Managing Clinical Risk*,
Health Informatics, DOI 10.1007/978-3-319-26612-1_9

implementing an SMS is that these cases are thought through at the time of defining the scope of the SMS and not when a particular issue arises.

9.1 Information Governance (IG) Violation

HIT systems frequently contain data which is personal and sensitive. Culturally we have come to expect that our clinical data will be kept secure and only be divulged to those individuals with a stake in managing our well-being. Exposing this confidential data to those whose interests are financial gain or sheer curiosity is generally considered unacceptable and indeed unlawful. Information Governance represents the preservation of information confidentiality – i.e. the absence of unauthorised disclosure of information [1].

Harm is defined in IEC 61508 as "physical injury or damage to the health of people either directly, or indirectly as a result of damage to property or to the environment" [2]. So, given this definition, does a scenario where IG is breached constitute a safety incident? For example, suppose a clinician becomes aware that a friend or relative (who is not under their care) has been admitted to hospital. The individual, out of curiosity, accesses and views the patient's electronic record. The question here is has the patient come to harm? One could argue that the patient has suffered psychological harm, but what happens if the information isn't divulged further and actually the patient never finds out? Suppose viewing the information was done not maliciously but with the very best of intentions, does this impact our view? One might argue that a breach of patient confidentiality only represents harm in the event that the information is used to discriminate against the individual such that they suffer financial or reputational loss.

Debate in this area soon becomes a quagmire of morality, ethics and law which from a practical perspective of CRM is best avoided. What is important is that:

- An organisation recognises that IG and its management is an important requirement in any HIT system.
- There should be formal processes in place for managing an approach to IG.
- Whether or not IG is managed under the safety umbrella or some other organisational process should be agreed and documented in the SMS.
- If one encounters a situation where there is conflict between protecting information and protecting patient safety a careful argument must be constructed. Many authorities would feel that patient safety takes priority.

9.2 Security Violation

A security violation is the situation where an individual deliberately gains access to a system when they are not authorised to do so – a potentially unlawful act in most countries. Most HIT systems allow users not only to view information but also to change it and initiate clinical activity which has the potential to be a source of harm. One might therefore conclude that gaps in a system's security model represent a hazard to safety and it is this inference which is worthy of further investigation. Another potential scenario is the execution of a denial of service attack – a situation where an individual bombards a system with so much electronic traffic that that it becomes non-performant or unavailable.

In each of these circumstances a deliberately malicious act would need to take place before harm would occur. The question is whether this should be managed under the scope of CRM or by some other means of assurance? This is a controversial topic and even safety standards differ in their views. As with IG, the message is similar – security of HIT systems is vital and represents a source of risk. Whether or not this is managed under the CRM framework or by some other means of assurance should be decided.

Finally note that on occasion, security measures can have a more direct relationship with safety. This typically occurs when security is overly restrictive and prevents a user from gaining access to the information he/she genuinely requires. This is discussed further in Sect. 6.5.5.

9.3 Operation Outside of the Intended Purpose

The need for a system to have an intended purpose is discussed in Sect. 11.1.2. From time to time systems are operated outside of their intended purpose either intentionally or unintentionally. Organisations or their users occasionally find innovative ways of reusing functionality that was meant to serve a different purpose. Users may well continue to operate the system according to their training and local policy only extrapolate the functionality to support clinical business processes for which the system was fundamentally not designed. Note that in some regulatory frameworks such as the EU Medical Device Directive, operating a device outside of its intended purpose has serious implications and advice should be sought from the manufacturer and regulator.

Where extension of use is conducted as part of a controlled roll out it should be subject to the usual change management procedures which will have been established in the in SMS. Essentially a set of safety activities will need to take place on the product to determine whether the new operating environment introduces any new hazards and the safety case updated accordingly. The analysis will need to be able to justify operating the system outside of the intended purpose and set out the evidence to support the claim that the risk remains acceptable.

The scenario one wants to avoid of course is where a system's purpose is expanded in an uncontrolled manner, without the knowledge of the project and CRM team. Preventing this is largely down to governance, leadership and diligence of stakeholders involved in the implementation and education of users in the system's boundaries.

9.4 Misuse

Misuse refers to inappropriate operation of a HIT system albeit within the product's intended purpose. Misuse exists on a spectrum:

- Mistakes – where users, acting with the best of intentions, perform an incorrect action. For example, accidentally ordering an inappropriate investigation.
- Violations – where users perform operations which contravene training and local policies. For example, deleting a duplicate patient record where local policy dictates that it should be merged.
- Gross violations – where a user violates the system to such an extent that the action could be regarded as reckless. For example, a junior nurse using a doctor's smartcard in order to prescribe a dangerous drug.
- Malicious act – where users deliberately perform actions to cause harm to a patient. For example, prescribing a medication with the intention of injury or terminating a patient's life.

Mistakes and violations can be relatively common and further discussion on human factors is provided in Sect. 5.3. Thankfully gross violations and malicious acts are rare but are still worthy of some thought. The question here is whether failing to prevent a user from acting maliciously or recklessly represents a hazard. Indeed, can any system truly guard against acts of this nature? Is it realistic for a HIT system to have the intelligence to reliably determine whether an action, in the context of a particular patient is clinically appropriate or reckless? Importantly should that intelligence fail and inappropriately deny a clinician access to an action they genuinely need to perform, that is likely to introduce more risk than the original issue.

Furthermore, perhaps this is one example where a comparison with the pre-electronic era is useful. We would not expect a telephone and a piece of paper to prevent an individual from acting in a malicious or reckless manner. Therefore when one introduces an electronic system which includes secure authentication and a full audit trail of user activity, does that system really now present a hazard by not preventing the same action? To take this discussion to its conclusion, one could argue that if an organisation relies upon a system to prevent its staff acting against their professional training then that in itself could constitute a far greater risk.

Attempting to assess malicious and reckless acts can quickly become a bottomless pit in terms of determining how one goes about introducing controls to guard against them. Organisations implementing a CRM system should therefore seriously consider whether these scenarios are within scope of CRM or simply exist as general healthcare delivery risks.

9.5 Harm to Individuals Other than Patients

So far we have considered only harm to patients. However when undertaking a hazard assessment, scenarios soon arise where the impact relates not to the patient but to a member of staff, their relatives or another patient.

For example, a system could fail to indicate that a patient suffered from an infectious condition. Clinicians may therefore fail to take appropriate measures to prevent themselves being infected. Does this constitute a hazard in scope of HIT CRM? Similarly, should a system fail to indicate that a patient has violent tendencies – does this represent a risk if the patient were to harm a member of staff or another patient?

Opinion on this varies and in many countries harm to staff is likely to be covered by other risk management frameworks such as that of the UK's Health and Safety Executive [3]. As with some of the issues presented above, it is important either way, to formally include or exclude harm to staff and other patients in developing the scope of the SMS.

9.6 Harm Arising from Patients Viewing Their Own Clinical Data

Increasingly healthcare organisations are making it easier to interact electronically with their patients. Securely opening up the patient record for their consumption has enormous benefits in better engaging patients in their care. Individuals often have difficulty recalling the details of a consultation with their healthcare provider after the event. By enabling access to their own notes patients not only have a single point of reference but they also benefit from being empowered to raise concerns when the information appears to be inaccurate or incomplete. Close examination of the processes however reveals that these features could potentially be a source of harm, again depending on how one chooses to define it. There are at least two distinct concerning scenarios.

Firstly one is required to examine the extended effects of incorrect data in the clinical record. For example, suppose that data erroneously gives the impression that a patient's diabetes was well-controlled when, in fact, this was not the case. Potentially the patient could change the manner in which they approach their condition, become lax in managing their diet or reduce their compliance with medication instructions. A record can be misleading if the underlying data is incorrect but also, and perhaps more likely, when the manner in which it is presented to the consumer is suboptimal. Patients are unlikely to be expert users of the system and translating the entire medical record into a form that facilitates safe patient interpretation is no easy task.

Secondly, thought should be given as to how support is provided for information which is released to the patient in electronic format. Suppose a laboratory test result is received by the system and is made automatically accessible to the patient. That simple data point could be life-changing for the individual concerned either appropriately or erroneously. A patient's understanding of the information intended for a healthcare professional is also likely to be limited. When someone relies on an Internet search engine to support interpretation of the information in the record this can quickly give rise to confusion and (often needlessly inflated) concern. Add to this the fact that society is increasingly making use of mobile technologies to consume data. We must therefore acknowledge the very real possibility that an individual could receive devastating news about their wellbeing not in the controlled environment of a physician's consulting room but on the 10:47 from London King's Cross.

9.7 Assessing Absent Functionality

A curious dilemma occasionally emerges when one is required to assess functionality which is not present. At a most basic level one might conclude that it is simply impossible to assess functionality which is not there – after all without any designs, architecture or user interface what is there to evaluate? But the situation is worthy of closer examination.

Firstly one must differentiate between functionality which has never been available and that which becomes unavailable in the live environment. Should a clinician rely upon a particular system, module or function to deliver safe care and that entity become unexpectedly unavailable that is clearly a hazard and is discussed in Chapter 7.

For functionality which has never been available in a system it is important to carefully think about whether its absence genuinely introduces risk or whether it is a case of our expectations as a user which are not being met. Suppose our organisation procures an Electronic Health Record system and transformation activities begin in order to migrate to new processes. It might be that we subsequently discover that the prescribing component has shortcomings which cannot be overcome and is selectively removed from scope. Whilst this may be a disappointment and preclude us from realising the full benefits of the system, it is almost certainly a mistake to frame that discontent as a clinical risk. After all it is likely that one is able to continue with existing manual business processes which have been in place for long periods of time.

One argument which is occasionally developed by healthcare organisations goes as follows:

1. We have a directive to effectively manage clinical risk and patient safety
2. The shortcomings of the system do not allow us to realise the benefits we were hoping to realise
3. Some of the benefits included improved patient safety, therefore
4. The absence of the required function and the system itself represents a risk and it must be unsafe.

It is the last step in this chain of logic which is subject to question. The reality is that whilst the system fails to control existing risk, in itself it does not introduce new hazards to the clinical environment therefore it cannot be unsafe (at least not on these grounds). The argument is usually borne out of frustration and with the very best of intentions. The real issue here is that these matters are best dealt with through an examination of procurement methodology, commercial arrangements, requirements definition, expectation setting and supplier engagement all of which are vitally important in a HIT project but are distinct from CRM.

At the other end of the spectrum is the scenario where we identify before go-live that functionality which we would typically rely upon to deliver care is missing. For example, suppose that same electronic prescribing system had functionality to prescribe a drug but not to discontinue it. Here the premise that absent functionality cannot introduce risk begins to fall down. So how does one build an argument around something that does not exist? The trick in this case is to frame the medication management business process as including both the prescribing and discontinuation steps. Discontinuation mitigates the risk that a drug is inappropriately administered for longer than it is clinically required. The absence of this control therefore leaves the hazard with an intolerable degree of risk and, as such, re-engineering or a change in project scope is required. A similar situation arises if a user is led to believe that that a piece of functionality or logic exists when in fact it doesn't. Through training and sound user interface design the scope of the system needs to be perfectly transparent.

One final scenario is where functionality is descoped close to go-live. This is not uncommon in HIT projects and can represent a genuine, although often unpalatable strategy for risk control (see Sect. 15.1.1). But the question arises as to whether removing functionality from scope itself introduces risk. This can usually be solved by analysing the dependencies which exist between the target functionality, the rest of the system and the supported clinical business processes. Continuing our example, suppose the pharmacy department is reliant upon the implementation of electronic prescribing to be able to support their day to day work. Descoping the module without their engagement could introduce significant risk late in the project. Similarly should users have received training on the functionality and perhaps begun to transform their clinical practice and processes, any descoping would need to be carefully managed and communicated.

In summary, the absence of functionality does not normally introduce clinical risk where the restrictions are a simple failure to meet one's expectations or realise benefits. On the other hand where the constraints are not fully understood, or the situation result in missing controls or key dependencies, absent functionality can contribute clinical risk.

9.8 Medico-legal Issues

Occasionally HIT systems support a healthcare organisation's medico-legal activities. SMS developers should consider whether failure of these systems constitutes harm. For example, most HIT systems contain audit trail functionality to maintain a record of which users undertook which actions and when. This information can be retrieved from the audit log should it be needed to support the investigation of complaints or legal action. If the system fails to maintain this information this may constitute a breach of data protection law but not necessarily harm in the context of CRM.

Similarly, HIT systems may support specific legal frameworks. For example a system may be used to maintain a record of which patients are currently detained against their will for psychiatric treatment. If the system were to indicate that a patient was subject to detention when in fact this was not the case they would be detained illegally, would this fall within scope of CRM?

9.9 Summary

- To notion of harm itself can be contentious so it is unsurprising that some scenarios are on the boundaries of CRM.
- The boundaries should be set early and formally documented in the SMS during its development rather than as challenges arise.
- In particular, issues of security and Information Governance are common in HIT systems and a decision needs to be taken early as to whether or not problems in this area are managed under the auspices of safety.
- Systems can only go so far in preventing deliberate or inadvertently reckless acts and the safety case should reflect this real-world limitation.
- Systems which are patient-facing rather than clinician focussed should not be automatically excluded from the risk assessment as they can still cause harm.

References

1. Avizienis A, Laprie C, Randell B. Fundamental concepts of dependability. Research Report No 1145. LAAS-CNRS. 2001 April.
2. International Electrotechnical Commission. International standard IEC 61508: functional safety of electrical/electronic/programmable electronic safety related systems. Geneva; 2000.
3. Health and Safety Executive. Controlling the risks in the workplace. [Online]. 2015 [cited 2015 July. Available from: http://www.hse.gov.uk/risk/controlling-risks.htm.

Chapter 10
Evidencing a Competent Team

10.1 The Need for Competency

The evaluation of clinical risk requires the careful application of judgement based on a sound knowledge of the system under consideration and the domain in which it is deployed. Similarly, the effective formulation and implementation of an SMS is a complex task which requires skill and expertise. It follows therefore that the accuracy of a CRM analysis is significantly dependent on the competency of the stakeholders involved. Those challenging the claims of a safety case might be quite justified in bringing into question its validity where the capabilities and proficiency of its authors is questionable. It is therefore wise for a rigorous CRM system to be backed up by pro-active evidence of a personnel competency assessment for project stakeholders.

The notion of competency as a requirement in CRM is well established in other industries such as nuclear power. In engineering the term SQEP (pronounced 'squep') is used to refer to those who are Suitably Qualified and Experienced Personnel. SQEP implies that an individual is a professionally qualified person who has several years as a practitioner, is registered in their area of expertise within the organisation and so whose judgement can be used to resolve a technical problem with some finality [1]. In many cases formally establishing SQEP in these industries is a fundamental pre-requisite to employment or allocation to a project.

To evidence competency it is rarely convincing to merely state an individual's job title or role in an organisation. Whilst one would hope that an institution of good standing would be discriminating in choosing whom it appoints this is hardly transparent evidence in the context of a rigorous safety case. Similarly, academic qualifications alone are a somewhat blunt tool in establishing ones skills and past experience.

In some cases, demonstrating competency may be a solid contractual or regulatory requirement. A number of standards explicitly demand that one or more key

A. Stavert-Dobson, *Health Information Systems: Managing Clinical Risk*, Health Informatics, DOI 10.1007/978-3-319-26612-1_10

stakeholders possess certain credentials or experience. For example UK standards ISB 0129 [2] and ISB 0160 [3] demand the nomination of a Clinical Safety Officer who is responsible, amongst other things, for approving all CRM deliverables. The standard requires the individual to be a "suitably qualified and experienced clinician" who must "hold a current registration with an appropriate professional body relevant to their training and experience." Such an obligation can be a challenge for smaller manufacturers who may not have a clinician on their payroll and instead rely on domain input from customers or other informal sources.

A standard or commercial contract might also have an explicit requirement for individuals to have experience of CRM specifically, again quoting ISB 0129/ISB 0160, "Personnel MUST have the knowledge, experience and competencies appropriate to undertaking the clinical risk management tasks assigned to them." Similarly ISO 14971 [4] specifies, "Persons performing risk management tasks shall have the knowledge and experience appropriate to the tasks assigned to them."

In these cases it may be necessary for individuals to source specific training, particularly where an organisation is undertaking an assessment for the first time. More experience practitioners should maintain records of their experience and training as a ready source of evidence.

Particular care should be taken where organisations out-source some risk management activities to third-party organisations. Establishing commercial and regulatory competency in these circumstances can be complex. For example, the EU Medical Device Directive is explicit in that the Manufacturer of the product bares full responsibility irrespective of any commercial arrangements they may choose to impose through commercial contracts. In choosing to involve a third-party one should consider whether the organisation has sufficient domain and product knowledge to be able to formulate a credible safety case in a manner that can be evidenced. Similarly, it would be wise to formally acknowledge that accountability remains with the primary organisation.

10.2 Competency Assessment Methods

10.2.1 Curriculum Vitae Repository

As with most concepts in risk management it is right to invest greater effort in evidencing competency where a higher degree of assurance rigour is required. For smaller projects and where the overall clinical risk of the system is limited, evidencing competency can be as simple as maintaining a repository of up to date Curriculum Vitae for the stakeholders involved in the project. The CVs should highlight experience relevant to CRM rather than be inferred from previous roles. Details of significant training, previous safety projects undertaken and technical or clinical experience all go a long way to justifying ones right to engage in or lead a CRM project.

10.2.2 Skill Mapping

One limitation of a CV archiving approach is that it fails to objectively evidence the skill mix that is typically needed for the assessment of larger systems. An experienced clinician is likely to be well versed in the intricacies of the clinical business processes but relying on that individual alone to assess the technical resilience of the solution might have shortcomings. An alternative, more rigorous approach uses skill mapping to determine which individuals are appropriate to occupy specific project roles. For example, a reasonable technique might be:

1. Determine, for a particular project, what roles need to be fulfilled within the CRM team to provide an appropriate skill mix. For example this might include a clinical domain expert, risk management representative, technical expert, system administrator, training lead, etc. Each role should be accompanied by a brief job description and areas of responsibility.
2. Ascertain the skills, experience and qualifications required to operate in that particular role. For example, the domain expert might need to have at least 5 years practical clinical experience and possess an affiliation with an appropriate professional body. The risk management representative should perhaps have received specific training and be able to evidence previously issued safety cases. Note that these criteria should be established in a vacuum and not be reverse engineered based on the characteristics of the individual whom one has in mind for the role.
3. Formally assess staff against all of the criteria established in the step above. This can be done with a simple scoring system or binary true/false prompts. The outcome of this will allow one to objectively demonstrate that certain individuals are capable of operating in particular roles but not in others.
4. Formally allocate individuals to roles within the CRM team and record that within the CRM project file.
5. Ensure that personnel revisit the skill assessment from time to time in order to reflect new skills learnt and to detect deskilling.

Where a particular individual fails to meet the requirements of a role, a training plan is needed, at least for those skills which can realistically be met through training. Whilst the training plan is being completed, it may be possible for the individual still to occupy the target role in the team but with additional supervision and mentoring in place.

A key element of developing a culture of safety is the need to identify and manage training requirements. Documenting deficiencies in knowledge and experience therefore represent a strength in a rigorous CRM process rather than a weakness. The training records themselves should form part of the safety management file demonstrating a mature and vigilant approach to establishing competency.

Evidencing that staff are competent in their role can be enhanced by formalising the materials used to train members of the CRM team. In practice these are likely to include:

- Relevant standards and publications
- Locally developed processes and policies
- User guides for the tools supporting the project
- Publically available teaching resources
- Best practice guides

Larger organisations are likely to want to develop their own CRM teaching materials in the form of presentations and speaker notes with specific objectives and learning outcomes. These facilitate a coherent and consistent resource which can be purposed and repurposed for promoting a safety culture more widely in an organisation. Evidence that individuals have consulted, worked-through and digested the training materials should be maintained in a manner acceptable to potential auditors. This can take the form a simple checklist with completion dates recorded for each member of staff.

10.3 Competency and Resource Management

One of the advantages of undertaking a formal competency assessment process is that potential single points of failure can be easily identified. For example, suppose a key contributor to the safety project is a clinician who has undergone risk management training. Should it be the case that an organisation only possesses a single individual with those qualifications, this may present a project risk should that individual leave the organisation or is required to focus on purely clinical duties. Without the necessary knowledge input the safety assessment is likely to be significantly delayed. In these circumstances it would be prudent to invest in training or recruiting additional staff members to facilitate smooth running of the project.

10.4 Summary

- A safety case is only as good as the individuals who constructed it. Any doubt in the reader's mind about the competency of its contributors is likely to undermine its conclusions.
- There is merit in proactively seeking to demonstrate competency through the use of either a Curriculum Vitae repository or more formal skill mapping.
- Ensuring that the clinical risk management team is competent to perform their role is a key function of senior management.
- Demonstrating competency is a key requirements of some risk management standards.

References

1. Wikipedia. [Online]. 2015 [cited 2015 March. Available from: http://en.wikipedia.org/wiki/SQEP.
2. Health and Social Care Information Centre. ISB0129. Clinical risk management: its application in the manufacture of health IT systems version 2. UK Information Standard Board for Health and Social Care; 2013.
3. Health and Social Care Information Centre. ISB0160. Clinical risk management: its application in the deployment and use of health IT. UK Information Standard Board for Health and Social Care; 2013.
4. International Organization for Standardization. EN ISO 14971:2012. Medical devices. Application of risk management to medical devices; 2012.

Part IV
Undertaking a Clinical Risk Management Project

Chapter 11
Planning a Clinical Risk Management Project

We have already established that the undertaking of a CRM project should be conducted in a controlled and logical manner. Importantly the project should be carried out not in isolation but as an integrated component of the general product lifecycle or implementation. For these reasons wider project plans should contain explicit references to the CRM activities especially in the Project Initiation Document (PID).

Most well-managed projects have embedded within them a series of gates or milestones. These events usually have defined inputs and outputs with explicit expectations that certain materials and decisions will have been completed by that point. The gates are included in the project plan – a document owned by the project manager. It is into these gates that the output from the CRM analysis will feed. In this way the CRM project team is tasked with producing a clear set of deliverables at specific points in the project – formally agreed material which other workstreams will consume and utilise in their own work. These dependencies are important to define at the start of a project so that if timescales are unexpectedly drawn out the impact on other stakeholders can be quickly ascertained.

One of the first tasks in creating an SMS is to agree the precise deliverables which will be produced during a CRM assessment of a product and thereby make up the CRM file. One should aim to create the same set of reports for each product in order to maintain consistency in approach. Working this documentation into the project plan will then become routine for project managers. There is no rule-book governing the deliverables which should be produced but a number of standards including ISB 0129/0160 [1, 2] suggest at least three distinct documents which have been found to be practical in both HIT CRM and other industries:

1. The CRM Plan – setting out the safety activities which one intends to conduct
2. The Hazard Register – describing the individual hazards, causes and controls associated with the product and its implementation
3. The Safety Case Report – setting out the argument and evidence to support the safety claims made about the system under consideration

A. Stavert-Dobson, *Health Information Systems: Managing Clinical Risk*, Health Informatics, DOI 10.1007/978-3-319-26612-1_11

Note that this is by no means a definitive list and in some cases it will be prudent to split out a component of a report into its own document. For example, suppose system testing activities complete well ahead of the issuing of the safety case. Stakeholders may benefit from reviewing the CRM analysis which has taken place during testing but have a need to do this long before the full safety case is ready. In this case the CRM project team might choose to develop and issue a specific CRM Test Report to better manage those dependencies.

11.1 The CRM Plan

The CRM Plan is typically the first formal safety deliverable issued during a project. It fundamentally sets out the basis for the rest of the CRM activities. Note that the plan is not necessarily the same as the process. The process and SMS are agreed at an organisational level as a standard template or menu of CRM activities. The CRM Plan specifies the next level of detail – how the process will be applied to the particular product under consideration.

A typical plan might contain:

1. An overview of the project and how the CRM activities relate to the wider project.
2. A definition of the system and analytical boundaries. Reference should be made to any safety requirements which have been created.
3. Which components or functionality is safety-related along with the appropriate rationale.
4. Reference to any existing CRM work on which the project is building.
5. Roles and responsibilities.
6. A reference to the acceptability criteria in the wider SMS.
7. An outline of the individual CRM activities including hazard identification, control option analysis, verification activities, etc. (note that the plan should focus on areas where there is an intention to vary the approach from what is documented in the SMS).
8. The expected deliverables which are intended to be produced and the dates on which they will be issued.

Note that in many cases it will be appropriate to reference material in the wider project (such as the project plan or PID) rather than re-iterating it specifically in the CRM Plan. This prevents having to maintain information in more than one place. The Plan may be supplemented by the gathering of a safety planning meeting or committee. Alternatively safety matters may exist on the agenda of other wider project planning meetings.

At the outset of developing the plan it should be recognised that projects change and mature during their lifecycle. As such, it can be wise to revisit and reissue the plan from time to time (although a radical update immediately before the safety case is validated against the plan may prove questionable). In particular one needs to take a view on whether a proposed change in the project has an impact on the general CRM approach. For example, should the project scope change in a way that more or less rigour is called for, the planned CRM activities may need to be adjusted. The plan should be redrafted in this case so that project stakeholders are made aware of the alteration in a controlled manner.

11.1.1 Defining the Project Boundaries

At an early stage in the project planning it is necessary to carefully define the scope of the system or module under examination. Limiting and articulating the scope is necessary to define the boundaries that have been applied to the analysis. More importantly, this formalises those system entities which have not been subject to analysis. The safety case will therefore say nothing about the clinical risk associated with those components outside of that defined boundary. By instituting boundaries early on in a project one is able to more accurately size the target and define the resources and timescales necessary to complete the task.

Defining the boundary can be complex and often more complex that would appear at first sight. Consideration should be given to the logical architecture, the physical structure, people, organisations, configuration and supported business processes. At the simplest level, it is customary to formally define the product under consideration. One should be sure of the name of the product, who manufactures it and, importantly, which version is being analysed. Many manufacturers will update their systems on a regular basis and it is quite possible that a CRM analysis can begin against a particular version only to discover that come go-live a more recent version is deployed. Careful planning, co-operation with the manufacturer and an understanding of the product roadmap are therefore essential.

A similar situation exists for configuration control. Modern HIT systems are highly configurable and can be modified beyond all recognition by either the manufacturer or system administrators without any change in the underlying code. Configuration sometimes represents a moving target and for the developer of the safety case, this can prove to be a challenge. Again, careful planning is the order of the day preferably through the implementation of a formal and documented change management process.

Beyond defining the product and configuration often the easiest place to start setting a boundary is with the logical system architecture. For example, the system may be modular with well-demarcated interfaces to other components. In this case, it is reasonable to apply a boundary around the module and either include or exclude the relevant interfaces. On the other hand, it is quite possible that a module is only ever deployed in conjunction with other modules and therefore this level of granularity may be inappropriate.

One should also consider the physical architectural components of the system. Would it be appropriate to undertake separate safety analyses on the processing, storage and network components of the system? For example, suppose a hospital network is an integral component of a dozen different systems all of which contribute to clinical care. Should there be a dedicated safety analysis of the network itself and/or should network issues be included as failure modes in the safety case for each individual system? There is no right or wrong answer, what is important is that boundaries are strategically defined and considered as part of a holistic system such that components do not 'fall through the gaps'.

For some systems, a geographical approach may be taken in defining scope. For example, one may need to consider the safety impact of a hospital's servers being destroyed in a fire or natural disaster. It would be sensible in this case to draw boundaries around those systems which are hosted together or have similar dependencies.

A key purpose of the safety case and the associated hazard log is to formally communicate those controls which other stakeholders are required to implement. In many cases, this translates into a natural boundary between the supplier of a HIT system and the healthcare organisation who implements it. Safety cases frequently demarcate the scope at this convenient level and also make it possible for the parties developing safety cases to mirror the commercial contract.

Sometime boundaries can better be defined by looking at commonality in the operation of the system, its goals or potential impacts. For example, a HIT system may provide a platform on which healthcare organisations can build data entry forms, workflow and reports. Each hospital department may utilise an entirely different set of configured screens and behaviours. The clinical business processes that the system supports and the potential impact of hazards are likely to be very different between dissimilar care settings. In turn, the system controls and indeed the degree of clinical risk could vary significantly. In this case it may be better to have a number of different safety cases. For example, one could develop a master safety case for the system generally and sub-cases or appendices that apply only to specific departments with specific configurations.

11.1.2 Intended Purpose

Intended purpose represents the use for which a product, process or service is intended according to the specifications, instructions and information provided by the manufacturer [3]. Hazards that arise from system operation essentially fall into two categories; those hazards that occur when the system is being used in the way in which it was intended and those which occur only when it is used outside of that intention. Section 9.4 discusses the practicalities of assessing hazards that arise from system misuse. However, one can only place hazards in this category after the system's intended purpose has been clearly defined.

Defining the intended purpose can be a challenge particularly for HIT systems which are highly configurable and can be implemented across different care settings. Manufacturers need to be cognisant of the various media which might be used for communicating the intended purpose, either explicitly or implicitly. For example, a typical HIT system will have marketing collateral, training material, online help, product labelling and regulatory declarations. It is essential that this material consistently and accurately communicates the intended purpose. It is therefore useful to have a single point of truth which articulates this important definition.

The CRM Plan should state or reference the agreed intended purpose and in this way contribute to the boundary setting of the CRM analysis and safety case. As a minimum a HIT system's intended purpose should include:

- An overview of the system functionality offered
- The clinical business processes and workflows that the system will support
- The environment and care setting in which it will operate
- The kinds of users who will typically make use of the system and their clinical roles
- The key clinical benefits
- Whether the system has the potential to directly or indirectly impact clinical care
- An overview of the technology utilised and the warranted environment

Over time there is always the opportunity for scope creep. The intended purpose may change or grow as the product is developed and healthcare organisations find increasingly innovative ways of harnessing a system's functionality. It is therefore important to monitor the currently defined intended purpose against its live operation and the product roadmap. Where the intended purpose changes, it is likely that the safety case will need to be revisited and the clinical risk reassessed.

11.1.3 System Dependencies

Many software systems have dependencies on other software components and whilst some of these may be well defined, others may remain a little nebulous. Either way the relationships need to be understood particularly where the availability of one system is critical for that of another. An interesting question also arises in thinking about how deep is one obliged to delve into a system's dependencies? For example, a HIT system may utilise the Microsoft. Net framework to provide a number of shared libraries and common functional components. These components will typically rely on the Microsoft Windows operating system, the operating system will have dependencies on various software drivers and so on.

Knowing where to draw the analytical boundary can be an awkward dilemma particularly when dependencies are ill-defined or are of unknown pedigree. Ultimately an informed judgement needs to be made. That decision should be primarily informed by the degree of potential clinical risk associated with the component in question. For example, suppose an organisation manufactures the software which resides in an infusion pump. Even the smallest malfunction could cause death or serious injury. It would therefore be perfectly reasonable to include the device's operating systems and drivers in the safety analysis. In contrast a system designed to support hospital bed management, running on a proprietary operating system and utilising a popular commercial database product is unlikely to justify this degree of analytical depth and rigour on its dependencies.

11.1.4 "Black Box" Analyses

When one draws a boundary around the system the general intention is to perform a safety analysis on the contents. Normally that would involve a deep investigation of the technology, its design, build and test and processes and procedures that have gone into its manufacture. However, in some cases, that information is simply not available to the party undertaking the analysis. There are a number of reasons for this but often it comes down to a lack of information sharing and the protection of intellectual property.

For example, a healthcare organisation may receive a system or component from a manufacturer who has chosen not to undertake (or communicate) a safety assessment. Where a product falls outside the scope of statutory regulation, this is not uncommon. However, the healthcare organisation may still wish to assure itself that the clinical risk associated with the product is appropriately managed. The healthcare organisation may approach the manufacturer and ask to see the software designs, understand how subject matter experts have influenced the product, assess the degree and outcome of testing activities and have a view of known live issues logged by other customers. The manufacturer may simply refuse to provide this information citing confidentiality, commercial constraints and preservation of intellectual property. This raises the question of how the healthcare organisation might then go about assuring the product in such a knowledge vacuum.

In these cases one may need to treat the system as the proverbial 'black box'. An entity whose inputs and outputs are broadly understood despite ignorance of what happens in between. Assuring the system will broadly focus on whether the outputs of the system can be predicted from the inputs through system testing against the product's requirements. Obtaining as much documentation as possible can clearly help in planning and scripting this kind of testing.

Softer forms of evidence may also contribute to a black box safety case. For example, if the system has previously been deployed to hundreds of similar healthcare organisations, has a long-standing industry track record this may give some confidence in assuring its operation. Whether or not this evidence is sufficient to mitigate the identified hazards will be dependent of the degree of initial clinical risk.

11.1.5 Retrospective Analyses

There are overwhelming advantages to conducting safety activities which are integrated into the product lifecycle from conception. In particular, the opportunity to mitigate identified hazards at the design stage can result in more robust controls and solid evidence. However, it is recognised that this approach is not always practical. For example, an organisation may have operated a HIT system for many years without a formal safety case ever being developed. There may well be a wealth of operational knowledge within the organisation and indeed strong evidence of its historic safe functioning. This knowledge can be leveraged to produce a justified safety case long after the system has gone live if required. One should however avoid the retrospective approach simply as a matter of convenience or as a result of oversight. Wherever a methodology integrated into development and implementation is feasible, the opportunity should be seized. Where this cannot be achieved the retrospective approach is still far better than continuing to operate without a safety case at all.

A key danger in the retrospective approach is to fall into the trap of conveniently building the safety argument around selective evidence which happens to be at hand. To do so runs the risk of forcing the safety case to articulate an artificially rosy picture of the world. It can be easy to assume that a much-loved system that has been in operation for many years must be intrinsically safe. That is not to say that it is unsafe but one should actively question whether the data which happens to be available can truly and objectively substantiate the safety case's claims. In some circumstances it may be appropriate to undertake the hazard assessment in a virtual vacuum of operational knowledge perhaps involving personnel who are somewhat removed from its day-to-day business. In this way appropriate controls can be developed and more objectively be tested to determine whether they can be truly validated.

Another shortcoming of the retrospective approach is the difficulty one encounters in developing new controls. For example, suppose a hazard has been found to be incompletely mitigated and further risk reduction activities are required. Changing the design of the system at such a late stage can be awkward and costly (note however that this in itself is most definitely not an argument for avoiding the development of a safety case in the first place!). By developing a clear, balanced and rational argument for a safety-related design change, a healthcare organisation is likely to be able to make a good case to a supplier to upgrade the system. In contrast, an ill-defined and articulated new requirement may sit on a supplier's 'nice to have' list for many years without its priority being fully understood.

In many cases, any new controls that are needed may be easier to manage if they are within the domain of the organisation developing the safety case. For example, a retrospective safety case might identify the need for improved training or the implementation of new operational policies. Whilst these measures may not be as effective as a design change, they may reduce the risk to an acceptable level until a design change can be implemented.

Organisations intending to construct a retrospective safety case should give careful consideration to the resources required to undertake it. There can be a temptation to assume that a retrospective analysis can be undertaken much more quickly or with a lesser degree of rigour. Mature projects may have a wealth of evidence but this evidence can take time to assimilate and rationalise – particularly should that evidence turn out to demonstrate contradictory findings. As with all safety analyses, the resources allocated to the project and the intended degree of rigour should be commensurate with the degree of expected clinical risk.

11.1.6 Managing Assumptions and Constraints

Unfortunately it is never possible to know everything about every system, particularly at the time that a safety assessment is undertaken. It is important to identify and carefully manage those things that we know we do not know – the so-called 'known unknowns'.

In the first instance, it might be possible during the analysis to make assumptions about those things that we do not know but can reasonably expect. Whilst one may argue that 'one should never assume' in the real world this is not always practical and in fact dealing with assumptions simply becomes part of the risk management process itself. However, it is dangerous to make an assumption and fail to declare it or, importantly, fail to test it out and monitor it. By stating an assumption up-front, it gives stakeholders an active opportunity to contest it early and guard against claims against safety case's validity later on. Similarly, declaring an assumption and failing to receive a challenge after solicitation suggests that other stakeholders at least feel that the assumption is reasonable.

Assumptions that are made at the time of carrying out a safety assessment may later be proven valid or invalid. It is therefore sensible (and indeed desirable) to revisit assumptions from time to time and either declare them as facts or revisit the safety case to investigate the effect of the assumption being invalid.

In some cases there may be cross-over between identified controls and assumptions, particularly where the controls are outside of an organisation's sphere of influence. For example, a manufacturer may assume that all users of the system have been appropriately trained by the healthcare organisation. A training control of this nature could probably offer some degree of mitigation to nearly every hazard. Stating this specifically against every hazard can quickly lead to a bulky and unmanageable hazard log. It may simply be better to make an assumption on the project generally that healthcare organisations will appropriately train their users. This could be re-enforced by having a single hazard describing the potential for users to be insufficiently trained and the training controls placed against this.

On occasion one encounters a situation where information is simply not known and no practicable assumptions can be made. In this case we may have an opportunity to undertake a part-assessment perhaps until a time when the information becomes unavailable. This is a perfectly reasonable approach so long as this limitation is formally documented as a constraint on the analysis and the project generally. It may even be appropriate to document which areas of the analysis have potential to change once more information does become available. Clearly it is important to monitor the provision of the information on which the safety case is dependent and build this into the project plan. If one believes that the information may not be available or forthcoming then this should represent a risk to the project and be documented and escalated accordingly.

If at the end of the analysis when go-live is nearing the constraints are still in place, the safety case is essentially incomplete and the situation needs careful management. A view should be taken on whether the missing information represents a potential source of new hazards or controls. If it is the case that controls cannot be validated without the information but the safety case still stands up despite this then it may be reasonable to forge ahead with the constraint in place. In other circumstances where key information is not forthcoming, the project may need to be re-evaluated.

11.1.7 CRM in Projects with an Agile approach

Traditionally software has been developed using what is often referred to as a waterfall model. The waterfall model is a sequential, plan-driven process where each stage of development is completed before the next commences. The approach applies simple order and structure to a project beginning with a clear set of exhaustive goals and requirements and finishing with a validated product. A key characteristic of the waterfall method is the creation of rich documentation at each stage of the process. The requirements are defined in their entirety, a design is generated to fulfil all the requirements, the system is built and then tested from start to finish. The SMS in turn defines activities which sit alongside each of the phases executed in tandem with the development project.

Although tried and tested, the traditional waterfall approach suffers from a number of shortcomings particularly when a need arises to revisit earlier phases. Central to the method is the capturing of exhaustive requirements early in the lifecycle. Should these foundation requirements be faulty or incomplete this can threaten the entire project late in the day. Similarly, should (or indeed when) defects are found during testing, re-work can sometimes be extensive and difficult to implement once the system has been constructed. As verification occurs separately from the software's creation, defects are essentially built into the system and can go undetected for long periods of time. Often the fix for a defect is a suboptimal workaround chosen not to maximise quality but to address the issue in a manner which has minimal impact on the project timelines. Should customer requirements change or the scope be modified during development, this can introduce a delay in delivery.

The limitations of the waterfall method were being realised in the 1990s. In 2001 the notion of the Agile methodology was famously coined at meeting of 17 free-thinkers at the Snowbird Ski Resort, Utah. Agile has grown into a collection of approaches with similar developmental characteristics but which conform to the Agile Manifesto [4]. The techniques have been widely adopted in software engineering and variations have been successfully applied by a number of high profile organisations including NASA.

The approach is based on the notion that, in a complex system, it is difficult or impossible to fully predict what the final product will look like at the start of the project. Instead, high level objectives are set and these are worked towards by executing a series of iterative cycles or 'sprints' typically lasting between 1 and 4 weeks. The goal is that each sprint is a self-contained development cycle in which a selection of product features, chosen in partnership with the customer, are turned into working piece of tested software. In this way, development is incremental and iterative focusing on those requirements which add the greatest value.

Rather than defects being detected after the completion of coding the intention is to minimise problems by addressing faults early, verifying the code as it is written rather than many months later in system testing. Often the technique makes use of 'pair programming' where two individuals take it in turns to write the system code whilst the other observes and checks the work in real-time. By quickly placing the output of each iteration in front of the customer, the work can quickly be validated and issues addressed. Even if the incremental output is considered a prototype its richness and stability provides for a far more meaningful user validation and acceptance compared to a drawing or static screenshot.

Many myths surround the agile method, a common one being that documentation is not required. Documentation still forms a key part of the overall project and of each sprint but it is undertaken only to a point where it is sufficient to meet the next set of objectives. The intention is to eliminate unnecessary documentation and instead to focus on what is needed in order to create the next functional iteration of the product. Another myth is that agile is akin to working without a plan, a technique which lacks rigour and robustness. In reality quite the contrary is true. Agile represents a collection of good practices executed with a great deal of discipline and control and designed to create incrementally verifiable artefacts. It is not that deep analysis is omitted but rather it is undertaken at a time when sufficient information is made available for the reasoning to be justified.

So how does an agile approach affect the process of carrying out a safety assessment? Firstly most standards in HIT, even those at the medical device end of the risk spectrum, stop short of specifying a precise developmental methodology. Whether a waterfall, agile or any other technique is employed is relatively unimportant when it comes to compliance. Many medical device manufactures successfully operate using an agile framework. In fact the process of code generation, fast verification and refactoring can be highly attractive where quality is an important objective. Some standards such as the UK's ISB0129 and ISB0160 [1, 2] are supported by specific guidance [5] on the use of agile in CRM.

Some adaptation of the assessment process however may be necessary in order to practically provide organised evidence for the safety case. Early in the project it can be helpful to record specific safety requirements alongside each of the product features on the to-do list or 'backlog'. As development happens quickly there can be minimal time to begin the safety analysis from scratch when a sprint commences. Instead the safety requirements need to be consumed into the process from the start ready to be turned into code. Some indication of the clinical risk associated with each feature is also a useful indicator as this can be used to set the benchmark for the risk involved (and therefore the rigour required) in implementing it. The risk is one factor which should be taken into account in selecting which features from the backlog will be selected for a particular iteration.

The argument set out in the safety case requires there to be coherence of user stories, requirements, code, testing, controls, etc. When documentation is at a minimum safety case developers can find it time-consuming to construct a logical argument and articulate the relevant evidence in a fast-moving environment. A useful tool in this case is the notion of traceability, the connection of related materials to demonstrate an unbroken chain of evidence and logic. For example, one might create traceability between a user story, a piece of code, a hazard, a test, a set controls, etc. In this way should a test fail, a requirement change or code fail to be written in time the impact on an iteration can easily be ascertained from the web of dependencies. In terms of the bigger picture, traceability provides some logical glue to determine the impact of each iteration on the product safety case and overall clinical risk.

In many ways each iteration should be considered a safety project in its own right given that the intention is to produce a piece of working code. Each cycle will need to include a safety assessment and conclusions to be drawn at the end of the sprint on whether the risk can be justified. Particular care needs to be taken to ensure that any new hazards or causes of existing hazards introduced in the iteration are also mitigated in that cycle. It would generally be difficult to justify introducing a significant risk in one iteration with a plan to only mitigate it in a later cycle, especially when the intended output is deployable code. The nature of agile is that requirements are fluid and changing and one could easily lose sight of an important control if it were put off until a later date. Again, the instigation of traceability will assist in ensuring that the right controls are included in the scope. Similarly one needs to be cognisant of the fact that whilst each iteration might have an acceptable risk profile that new hazards can emerge once components are assembled.

In an agile method, careful thought needs to be given as to when a safety case is issued and risk acceptability re-evaluated. Where software increments are put into live service there will be a need for a manufacturer to issue an updated safety case to the healthcare organisation in good time for them to consume the information and take any necessary actions. The safety case needs to take account not just of the new functionality introduced in the iteration but also any impact on existing hazards identified to date from previous cycles. Similarly where any features could not be completed in an iteration, an impact assessment on the safety profile is warranted. Agile may be light on documentation but the safety case itself is perhaps not a good target to skimp on content.

11.1.8 Managing Project Change During Development

All projects change and mature over time, especially in IT. Often the longer the project the more opportunity there is for people to change their minds, there be a shift in legislation or alterations in best practice. Software development processes vary significantly and some are more tolerant of change than others. As described above, an agile approach actively embraces change but larger more complex project are often less flexible. Project managers need to expect and plan for change and the software development lifecycle needs to evolve in such a way that change can be effected safely. Without reflecting change in the safety case its arguments can quickly become invalid.

At least two key changes in a project are significant; the addition of new safety related functionality to the product and the introduction of new controls. Where new functionality is brought into scope it needs to be subject to a safety assessment in the usual way – the fact that it is introduced into an existing product development phase is certainly no mitigator of risk. There can sometimes be a temptation to miss out assurance steps once a project reaches a certain level of maturity but this is usually difficult to justify. One must also consider whether introducing new functionality has an impact on the rest of the system particularly where dependencies are involved or where already assured components require a degree of disassembly. Where new controls are introduced into scope this is likely to be a welcome change from a safety perspective and the resulting impact on clinical risk should be brought out in the safety case. Of course controls themselves have the potential to mitigate one risk but introduce another so these too should be subject to a safety assessment.

Dealing with change during development has much crossover with changes implemented in live service. Section 20.1 discusses the management of change in more detail.

11.2 The Hazard Register

The hazard register is a key component of the SMS and safety case. Indeed one might argue that if only a single CRM activity were to be undertaken and documented it might be the development of a hazard register. Note that the term hazard log (not to be confused with an issue log) is also used commonly and this is essentially interchangeable with hazard register.

The hazard register represents a knowledge base, a structured and carefully considered body of work which spells out the potential sources of harm within the assessment boundary. Each hazard will typically be supplemented by further explanatory information; the precise triggering factors, means of control, potential impact and of course the evaluated clinical risk based on severity and likelihood. Through the hazard register essentially one is able to view the underlying data on which the rest of the safety case will be built. In this way the hazard register forms the basis of the argument which will be made in the safety case as well as evidence that a structured hazard identification process has taken place. The hazard register is also likely to be the primary means by which risk is communicated at a detailed level to other stakeholders – possibly across organisational or departmental boundaries.

In most projects the hazard register will continue to evolve throughout the product development lifecycle and into service as more is learnt about the system and the way it is operated. There is therefore no concept of a completed hazard register. Instead one may choose to take a snapshot to represent current knowledge and support any safety claims being made about the system at that point in time (for example at system go-live). That is not to say that a hazard register should be managed in a free for all or chaotic manner. Changes need to be subject to careful version control with each modification being scrutinised for its impact on existing material.

So what does the safety case have which the hazard register does not? Indeed why should one even bother to go to the effort of developing a safety case when such rich information is contained in the hazard register? What the hazard register lacks is narrative. In the same way that a series of blood test results provides us with clinical data, we still require a means of documenting our thought processes and conclusions we draw from that data. The safety case leverages the hazard register to develop a collective and comparative examination of the potential sources of harm culminating in a series of claims about the system's safety position.

There is another key practical difference between the hazard register and safety case. The moment material is placed into the hazard register it becomes useful. To a project stakeholder the detection of a hazard, understanding its impact or establishing a control is helpful and practical information. We may be able to begin acting on this, validating the design, authoring test scripts or consulting on a new operating policy. It is never too early to begin mitigating risk and populating the hazard register represents a quick win. Its living-breathing nature allows us to both develop it and consume it at the same time. In this way it is possible to gradually chip away at managing the risk tackling issues as they are recognised and gathering evidence of one's mitigation activities. The safety case on the other hand represents a summary view, the position when sufficient information has been gathered to formulate and evidence the safety claims. Both the hazard register and safety case have roles to play in CRM and it is useful at the time of planning a project to agree milestones when each will be produced to greatest effect.

Chapters 12 and 13 discuss the practical steps in formulating the structure of the hazard register and then populating it with material for a particular product assessment.

11.3 The Safety Case

11.3.1 What Is a Safety Case?

The safety case provides a structured argument, supported by a body of evidence, that provides a compelling, comprehensible and valid case that a system is acceptably safe for a given application in a given context [6]. Let's break that definition down into its component parts.

Structured Argument

The safety case is not a dashboard of data or a set of performance metrics. Its depth and interpretation is much greater and the foundation of the safety case is the concept of a structured argument. The articulation of hazard's causes and, importantly, the controls provides a rounded rationale from which any conclusions relating to overall degree of clinical risk will be drawn. The argument will be underpinned by the content of the hazard register and will be developed by supporting text in the safety case to describe the general themes, potential impacts and significant controls. Without argument ones runs the risk of trying to validate the safety position based on disparate and disconnected information sources. This can result in a safety case which is unfounded, piecemeal, illogical and incomplete.

The importance of structure in the safety case should not be overlooked and it is a characteristic which brings at least a couple of significant benefits. Firstly, structure makes the safety case readable and comprehensible to casual as well as expert readers. In the same way that a novel develops setting, story and characters, the safety case should flow. It should have a clear beginning, middle and end gradually building upon facts which are described and validated through the application of logical thought.

Secondly, structure provides the reader with a means of evaluating completeness. The goal-based assessment methodology and a number of common risk assessment techniques can occasionally be criticised for their inability to clearly demonstrate completeness. A convincing safety argument may be set out based on the evident facts but how can one be reassured that all credible hazards have been considered? The application of structure to an argument facilitates the reader in determining this and perhaps invites their welcome criticism.

Body of Evidence

If a safety case built without argument is unfounded then a safety case built without evidence is invalid or incomplete. One can make claims that risk has been appropriately mitigated but without evidence, how can one be convinced that those claims hold weight, that due diligence has been applied and that the controls set out in the hazard register are effective? Evidence should underpin all parts of the safety case and the application of formal structure and rigorous technique should facilitate evidence gathering and presentation. At the simplest level, it should be possible to take any control in the hazard register and draw a line to a documented activity which demonstrates its implementation. The granularity with which this is done will depend on the required degree of rigour.

Compelling, Comprehensible and Valid

Most HIT safety cases would not typically stand up to the intense rigour of the scientific method. The safety case is not a mathematical proof. Rather it should present the rationale for a reasonably educated reader, through the careful application of logic, to draw justifiable conclusions from the evidence presented. The technique is sometimes referred to as inductive reasoning and is the basis of many scientific theories.

A compelling case is developed through the rigorous application of process against a target which is appropriately granular and well-defined. Doing so establishes completeness with minimal room for hazards to fall between the cracks and remain hidden. A comprehensible safety case summarises the analysis in a concise and structured manner leading the reader through to logical conclusions based on established facts. Validity comes from the gathering of objective evidence and demonstrating that the evidence is complete. Validity builds confidence in the safety case and justifies the claims made.

Acceptably Safe

The acceptability of risk determines the extent to which we are prepared to tolerate risk. This has been discussed in detail in Chap. 3.

Given Application and Context

The risk associated with a HIT system is likely to be determined by the context in which it is operated. In other words, the clinical risk might be acceptable when the software is utilised to support care in one particular setting but, even with the same controls in place, might be unacceptable in a different setting.

Keeping control of this is dependent on a number of different factors which include the manufacturer's tight definition of the application's intended purpose and the healthcare organisation's ability to roll out the system in a coordinated and predictable manner.

11.3.2 Benefits of the Safety Case

Safety cases are gradually being adopted in the HIT domain especially in the UK in order to comply with safety standards ISB 0129/0160 [7]. Outside the UK the term 'safety case' is not widely used [8] but nevertheless reports which achieve the same purpose are becoming increasingly prevalent and may take other names such as a Safety Analysis Report. The safety case and its development has a number of benefits to manufacturers, healthcare organisations and ultimately to patients. These can be broadly categorised as [9]:

- Systematic thinking
- Integration of evidence sources
- Aiding communication amongst stakeholders
- Making the implicit, explicit

The safety case neatly packages a summary of the clinical risk in a single place. The input material for building a safety case is often diverse and wide ranging. The safety argument may be built from system requirements, domain and problem descriptions, user stories, design and architectural material, etc. Similarly evidence is likely to come from test reports, training material, policy and operational documents, etc. Only the safety case draws this rich source of information together in a manner that provides a comprehensive view of the clinical risk. Without the safety case, each stakeholder would be required to source and consider each piece of material individually and draw their own, and perhaps differing conclusions. In this way, the safety case serves a role as a record of engineering, a single place where the good work that has been done to reduce risk is succinctly and comprehensively set out.

A typical HIT project might involve hundreds of different individuals each with their own responsibilities and areas of interest. Frequently the controls required to mitigate the clinical risk will need to be implemented by many different stakeholders often across organisational boundaries. Ensuring that each control is examined can be a complex task and the development of the safety case allows there to be a single point of truth for what needs to be done and by whom.

One might be lead to believe that the development of the safety case is simply a means to evaluate and document the risk. However, those who have undertaken this kind of project quickly realise that the value goes far beyond a simple sterile exercise in evidence gathering. The development and articulation of the safety argument spawns questions about the system itself, questions which may not have previously been pondered. This can expose flaws in the safety argument which require careful thought and potentially further investigation and mitigation. The end position is a reduction and better understanding of the risk. In many ways the real value of the safety case may come from the process of its development rather than the end product [9]. In this way, the safety case itself whilst not intrinsically a direct risk control, is likely to induce behaviours which ultimately result in risk mitigation.

There are at least of a couple of other subtle benefits of the safety case. For manufacturers, products which are delivered along with a well thought-out safety case create a perception of quality, attention to detail and responsibility. The value which this adds to a product is very real and can generate a competitive edge over products from alternative suppliers. Manufacturers are quite justified in marketing the fact that a product has been subject to a safety assessment particularly where there is a statutory or regulatory requirement to comply with a standard.

There have been few legal tests but in the event that an organisation finds itself subject to a challenge, the presence of an SMS, safety policy and safety case helps to demonstrate adherence to good practice and the regard in which an organisation holds safety and risk management. Certainly establishing that a product is acceptably safe prior to any legal challenge arising is likely to carry more weight in court than any retrospective analysis conducted after a complaint has been lodged.

11.3.3 Potential Shortcomings of the Safety Case

When used to accurately summarise risk, communicate controls and contribute to a safety culture, the safety case is an invaluable tool in assuring a HIT system. However, the value of the safety case can quickly be eroded should it be seen as an administrative exercise or a simple means to an end to satisfy management, a regulator or contractual requirement. All too often the potential hazards associated with a system are only thought about when development of the safety case begins. This means there are dangers when the safety case is developed as an after-thought, where existing activities and evidence are conveniently reframed as risk mitigation and little or no attempt is made to integrate safety management into the underlying business processes.

Occasionally a safety case can serve as an attempt to simply shift the risk onto individuals further down the food chain without actually providing them with any useful risk mitigating strategies. A solid safety case should contain explicit controls which require individuals or organisations to take action; to test part of the system, to implement a policy, to operate the system in specific ways or adopt certain workflows or business processes. These controls are only likely to be effective if they are meaningful, relevant, specific and credible. From time to time one encounters a vague safety case which simply calls for individuals to be 'careful', 'appropriately trained' or 'professional'. In practice, this is unlikely to effect any significant change and undermines the true purpose of the safety case.

Finally, there is no doubt that the careful authoring of the safety case requires significant resources especially when processes are immature and/or a product is being assessed for the first time. However a well-designed process can make use of existing materials and activities. This, combined with the resulting improvement in quality which the analysis will drive, should provide a good investment when one considers the bigger picture.

11.4 Summary

- CRM projects need to be carefully planned in line with the development or implementation project. Timelines, dependencies and deliverables need to be agreed between stakeholders and especially across organisational boundaries.
- The deliverables produced during the assessment of a product should include a CRM Plan, hazard register and safety case.
- The CRM Plan should state the intended purpose of the product, define the boundaries of the assessment, state any assumptions and summarise the overall approach for the product under examination.
- The hazard register contains the fundamental building blocks of the safety argument and acts as an on-going repository of CRM knowledge for stakeholders to consume.
- The safety case is built on argument and supporting evidence and its creation helps organisations to better understand and therefore mitigate clinical risk.

References

1. Health and Social Care Information Centre. ISB0129. Clinical risk management: its application in the manufacture of health IT systems version 2. UK Information Standard Board for Health and Social Care. London. 2013.

2. Health and Social Care Information Centre. ISB0160. Clinical risk management: its application in the deployment and use of health IT. UK Information Standard Board for Health and Social Care. London. 2013.
3. International Organization for Standardization. EN ISO 14971:2012. Medical devices. Application of risk management to medical devices. Geneva; 2012.
4. The Agile Manifesto. [Online]. 2015 [cited Jul 2015. Available from: HYPERLINK http://www.agilemanifesto.org/, http://www.agilemanifesto.org/.
5. Health and Social Care Information Centre. Clinical risk management: agile development implementation guidance. Health and Social Care Information Centre. Leeds, UK; 2013.
6. Menon C, Hawkins R, McDermid J. Defence Standard 00-56 Issue 4: towards evidence-based safety standards. Safety-critical systems: problems, process and practice. York, UK; 2009. p. 223–43.
7. Despotou, G, Kelly, T, White, S & Ryan, M 2012, 'Introducing safety cases for health IT'. in Software Engineering in Health Care (SEHC), 2012 4th International Workshop on. pp. 44–50, 4th International Workshop on Software Engineering in Health Care, Zurich, Switzerland, 4–5 June.
8. Maguire R. Safety cases and safety reports. Meaning, motivation and management. Ashgate. ISBN: 9780754646495. Aldershot, UK; 2006.
9. Cleland G, Sujan M, Habli I, Medhurst J. Using safety cases in industry and healthcare. A pragmatic review of the use of safety cases in safety-critical industries – lessons and prerequisites for their application in healthcare. The Health Foundation. London, UK; 2012.

Chapter 12
Structuring the Hazard Register

Section 2.6 discussed the concept of the hazard, one of the fundamental building blocks of the hazard register. We turn our attention now to some of the other concepts; impacts, causes and controls.

12.1 Impacts of Hazards

A true hazard can only exist if it is possible to define a tangible and credible impact which results from the hazard being realised. The primary purpose of establishing the impact is to set the rationale for the estimation of severity – one of the two components of clinical risk. As we have previously discussed, the impact of a hazard should be expressed in terms of the harm it may cause on the patient rather than harm done to the system, its hardware or users. By focussing on the patient this ensures that the hazard and its impacts work together to describe a comprehensible and contextualised story from the trigger event to harm.

A hazard may have one or more impacts and each of these should be documented in some detail. It is reasonable in most cases to focus on the impacts which would be typical rather than the rare or incredulous. Note that in some cases it may become apparent that several hazards share the same impacts and some thought may need to be given as to how this is articulated. Duplicating identical pieces of text across multiple hazards becomes cumbersome and difficult to maintain. A better strategy is to detail commonly shared impacts elsewhere such as in the body of the safety case rather than in the hazard register. Alternatively it can be worth considering whether hazards which share identical outcomes are in fact flavours of a single hazard and could be better expressed in this way.

An example of a hazard and its associated impacts are shown below (Table 12.1).

A. Stavert-Dobson, *Health Information Systems: Managing Clinical Risk*,
Health Informatics, DOI 10.1007/978-3-319-26612-1_12

Table 12.1 Example hazard and its impacts

Hazard	Impacts
A member of staff could be allocated to the wrong clinical team on the system or to no team at all	Investigation results placed by the staff member could be sent to the wrong team and therefore not be available to support a clinician's decision making
	Referrals made to the individual might not reach them and instead go to the incorrect team. A patient could be inappropriately denied access to care
	Medications prescribed by the individual requiring authorisation might go to the wrong team delaying their approval and the provision of care
	Access to some functionality or information could be inappropriately denied

12.2 Causes of Hazards

Causes represent system characteristics or human behaviours which have the potential to trigger a hazard. For example:

- A power outage could cause total system unavailability
- Loss of connectivity could render a mobile user unable to log on
- A screen showing a long, unordered list of key clinical events could make it difficult or impossible to interpret

Causes should be credible and realistic and be worded in such a way that an educated casual reader is able to place the cause in the storyboard from incident to harm. Causes are distinct from hazards in that they don't represent the final situation that would lead to harm but rather raise the possibility that, given further prevailing conditions, an adverse event will occur.

Most hazards are triggered by a number of causes. If each hazard is only associated with a single cause one must consider whether the hazards are too specific and, in fact, represent causes of a higher level hazard. Similarly, hazards with dozens of causes might be too generic and become difficult to maintain in the long term. Often this balancing act needs to be revisited as the hazard register grows and content is merged or divided.

An example of a hazard and some associated causes are shown below (Table 12.2).

So how does one manage hazards where some causes are at a fine level of detail (perhaps referring to the behaviour of single button on a particular screen) and other causes of the same hazard are at a much higher level (perhaps unavailability of large pieces of functionality)? This opens up the thorny subject of granularity – just how specific or general should one's hazards and causes be? Get this wrong and you either end up with a hazard register which is too brief to be useful or too long to be navigable.

One solution is to maintain hazards at a consistent and generalised level of detail whilst allowing the causes to vary. In this way causes act as a granularity buffer; absorbing flexibility whilst forging a consistent level of detail in the rest of the hazard register. In most cases, having a mix of very detailed and specific causes and more general higher level causes does not lead to a problem even if they are causes

Table 12.2 Example of a hazard and its causes

Hazard	Causes
A member of staff could be allocated to the wrong clinical team or to no team at all	An administrative user responsible for allocating staff members to teams could select the wrong team
	Team names could be configured in a way that their description is confusing or ambiguous
	The relationship between staff and teams could become corrupt in the database
	Organisational data from the legacy system could be transformed incorrectly when imported into the application
	There could be a delay in updating the system when a member of staff moves from one team to another or when a new member of staff joins

Table 12.3 Example of a cause which is specific to a particular design issue

Hazard	Causes
A member of staff could be allocated to the wrong clinical team or to no team at all	The staff-team allocation screen shows only the staff member's surname and initial. An administrative user could misidentify the staff member where there are two or more users with the name surname and initial

of the same hazard. In fact, combinations of this sort can be beneficial. Very specific causes can highlight important system behaviours and are useful for spawning actions which need to be taken reduce the risk. In contrast wider, higher level causes can act as a catch-all for more general issues. In this way, hazards can be triggered by any number of related or unrelated causes, specific or generic, without adversely impacting the rest of the structure of the hazard register. In some cases, causes may represent a subset or superset of other causes belonging to the same hazard. However, documenting these as individual causes of the same hazard does not seem to have any deleterious effects.

For example, one could add the following cause to the scenario above (Table 12.3):

One should also consider at the time of authoring a cause whether it is described in a way that can realistically be controlled. Woolly, non-specific causes can be difficult to mitigate and even more difficult to evidence. Thus causes should be specific, to the point and facilitate control development. Of course if the cause is perfectly reasonable but controls still cannot be found, one has a potential safety issue that needs investigating further.

12.3 Controls

Controls are entities which attempt to mitigate risk and arguably form the most important component of the hazard register. Without documented controls, one would be faced with a one-sided safety argument for which the risk would be difficult to justify.

From time to time one hears the following logic: "I have identified a hazard – it has a number of strong controls therefore the hazard no longer exists". If one were to follow this to its logical conclusion the result would be a hazard register which only contains unmitigated hazards – perhaps something which would look like an issue log or fault log. As a rule, the presence of a control has no impact on the existence of a hazard or cause. The hazard is still there but happens to be kept in check by the presence of controls to an extent that the residual risk is now acceptable. Of course in the future those controls may or may not remain effective. By continuing to identify and document mitigated causes one has a handle on which to monitor risk going forwards. This characteristic of the hazard register turns it into a very useful tool indeed.

Interestingly, controls can come in two flavours depending on which part of the storyboard they interact with. The distinction between the two can be neatly articulated in a bow-tie diagram as shown below (Fig. 12.1).

Firstly, controls can act as barriers between a cause and a hazard – these are often referred to as proactive controls. An effective pro-active control will decrease the likelihood of a cause triggering a hazard. For example, a data entry form might be engineered to check if all key fields have been completed before allowing the user to continue.

Secondly, one may be able to mitigate the impact of the hazard by taking action after the harm has occurred – these can be referred to as reactive controls. In this way, one may be able to reduce the severity of the incident once it has been triggered. For example, suppose an electronic prescribing error results in a patient receiving an excessive dose of a drug. It may be possible to manage this scenario by monitoring the patient more closely or by administering another substance to counteract the pharmacological effect of the original drug. These are all valid

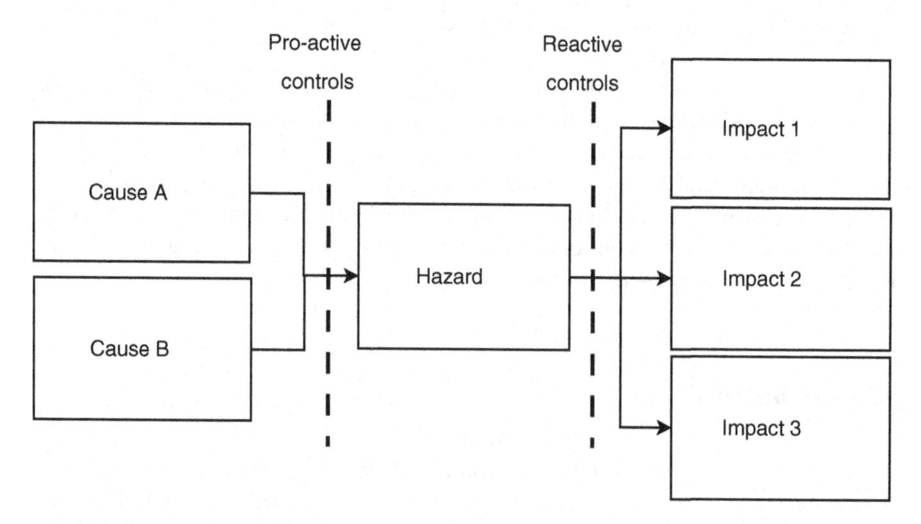

Fig. 12.1 Bow-tie diagram showing how causes, hazards, impacts and controls relate

controls in the wider picture of clinical medicine. However, one must question whether recording these kinds of controls is a useful exercise in the context of a HIT hazard register and safety project.

Reactive mitigation has different characteristics to proactive mitigation, especially in the HIT CRM space. A key distinction is that reactive controls nearly always rest entirely in the clinical domain and without direct involvement of the HIT system itself. These controls often reflect good clinical practice and, whilst vital to the delivery of care, the hazard register is arguably not the place to document these activities. In many cases reactive 'mopping up' controls contribute little to our strategy of preventing harm in the first place. Thus, the vast majority of controls which reside in the hazard register will usually be proactive. In this way, the controls will decrease the likelihood of harm occurring but without significant impact on the severity.

An example of controls linked to a single cause is shown below (Table 12.4):

12.4 Bringing the Structure Together

So far we have looked at the nature of impacts, hazards, causes and controls but we need to consider how these concepts hang together. In particular, how can we use these concepts to construct and articulate knowledge in the form of a hazard register?

The hazard, its causes and impacts are fairly straight-forward to model. The hazard is the starting point and will essentially act as the container – it will have a

Table 12.4 Example of a series of controls intended (or assumed) to reduce the chances of a hazard being triggered

Hazard	Cause	Controls
A member of staff could be allocated to the wrong clinical team or to no team at all	An administrative user responsible for allocating staff members to teams could select the wrong team	The system displays a confirmation screen before the allocation is made clearly showing the full details of the selected team and staff member
		The system training course will discuss the risk in attributing a staff member to the wrong team and explain that administrative users should use the confirmation screen to validate their actions
		The system will only offer for selection teams which are currently active and will not show those which have been retired
		The system contains a report which will print out the entire team structure. Administrative users should examine and validate the structure using this report after allocating a batch of staff members to teams

number of properties such as a code, title, description, etc. Next it should be possible to construct a series of potential impacts for the hazard and then the causes which may lead to the hazard being triggered.

Finally one needs to think through how controls are articulated in this structure and this is where the picture can become more complex. Are controls associated with individual causes, with the hazard itself or even the impacts? There is no right or wrong answer here and each approach has its merits. As previously discussed, controls which are linked directly to impacts tend to be of a reactive nature. A HIT hazard register will typically focus on the proactive so this linkage is often not terribly useful.

Recording controls directly against the hazard is fairly straight-forward in terms of structure. The result is that each hazard has a set of causes, a set of controls and a set of impacts all independent of each other but linked through a common hazard. Many hazard registers are constructed in just this way. This is a neat and concise approach but not without its limitations (Fig. 12.2).

The difficulty here is that without a logical relationship between the causes and controls, how can one ascertain that each cause has been appropriately controlled? Ultimately, the reader is forced to loosely evaluate the set of causes against the controls and make a judgement on whether or not that represents a complete analysis. This can at times lead to some subjectivity with an opportunity for critics to 'poke holes' in the safety argument.

An alternative approach is to associate each individual cause with one or more controls. In this way, full traceability is established. The reader is able to determine at a glance whether there are any causes without controls and similarly, should existing controls fail to be effective it is simple to determine which cause could then trigger the hazard. The approach brings an additional layer of traceability to maintain but the strong logical relationship between causes and controls can provide significant benefits (Fig. 12.3).

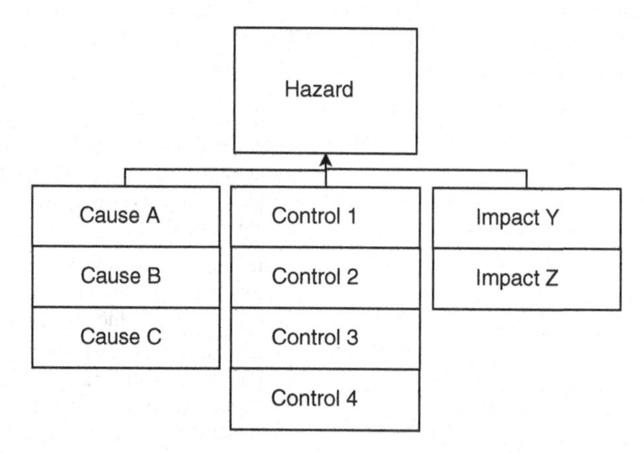

Fig. 12.2 Hazard register structure with each hazard having independent causes, controls and impacts

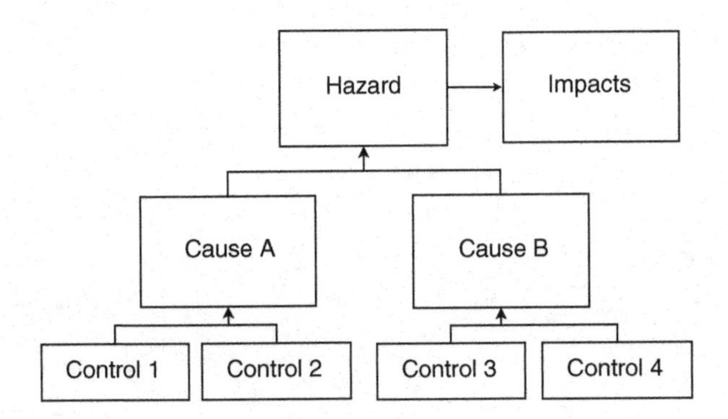

Fig. 12.3 Hierarchical hazard register structure

One shortcoming of this method is that should a single control mitigate more than one cause, it may be necessary to create several instances of the control, one against each relevant cause. When this occurs, it is usually better to do some restructuring of the granularity to prevent overly cumbersome repetition. If there are high level controls which essentially mitigate every cause (or even every hazard) one might be able to find an alternative home for this material, for example, a documented assumption against the entire project.

The latter approach works well in HIT CRM and can easily be expressed as a three-tier, one-to-many hierarchy; each hazard has multiple causes, each cause has multiple controls. Later in the analysis, each control will be evidenced (although this might not be set out in the hazard register itself).

Whichever approach is taken, it is important to decide on a hazard register structure and stick to it, preferably across all systems being assessed under the SMS. This brings consistency of approach, familiarity and the ability to compare (and sometimes transfer) material from one project to another. A template for the hazard register structure should be formulated and be documented as part of the SMS along with a brief description of how it is implemented.

12.5 Summary

- Hazard registers can be large and contain a significant volume of data. For it all to make sense it needs to be structured in a logical and consistent way.
- Hazards, causes and impacts can be arrange in a simple 'bow-tie' structure where each hazards is associated with multiple causes and impacts.
- Controls can be proactive or reactive but in CRM, where the aim is to intervene early in the chain of events, proactive controls should be the focus of the analysis.
- By linking causes to controls it makes it easier to work out where there might be potential gaps and examine the consequences of a particular control being ineffective.

Chapter 13
Populating the Hazard Register

13.1 The Use of Formal Methods

Once the structure of the hazard register is decided upon, at some point it will be necessary to populate it with content for a particular CRM project. A methodology for tackling this needs to be ascertained and there are a number of well-established techniques for driving out hazards. In the main, these have originated from engineering domains rather than healthcare specifically. Each has its pros and cons and there are benefits to employing different techniques for different parts of the system depending on the degree of risk, rigour, granularity of analysis required and the precise subject matter. Formal methods include:

- Structured What-If Technique (SWIFT)
- Failure Mode and Effects Analysis (FMEA)
- Hazard and Operability Study (HAZOP)
- Fault Tree Analysis (FTA)
- Event Tree Analysis (ETA)

Some of these methods can be difficult to apply in a HIT setting. For example, FTA and ETA use diagrammatic methods to link events and behaviours to hazards. However, for the techniques to work there generally has to be a clear relationship between cause and effect in the system, i.e. if A occurs and B occurs then C will occur. HIT solutions tend to be highly complex systems and the occurrence of harm is nearly always, at least to some degree, influenced by the system's human operators. Users are notoriously unpredictable in their actions and a system behaviour which might contribute to harm for one user may not for another. Thus the cause-effect relationship becomes loose and some systematic methods difficult to apply.

It should be remembered though that this limitation does not necessarily hold true for all parts of the system. Some components may exhibit a clear cause and effect relationship, particularly components that do not require user interaction; for example messaging between two or more systems or the execution of automated

A. Stavert-Dobson, *Health Information Systems: Managing Clinical Risk*,
Health Informatics, DOI 10.1007/978-3-319-26612-1_13

scripts. In these areas tools such as FTA and ETA may prove to be highly effective when expertly applied.

Another criticism of some systematic methods is their focus on low level system behaviour. Some techniques involve a detailed analysis of processes, sub-processes, components and subcomponents gradually building up the model until a view of the overall risk can be ascertained. These bottom-up methods are very sensitive at identifying potential hazards but in applying them one can quickly be consumed in death by detail. One of the foundations of risk management is the notion that the depth of the assessment and the effort applied should be commensurate with the degree of risk. The blind application of some systematic methods can undermine this principle and result in an over-complicated and time consuming analysis disproportionate to the degree of risk.

13.2 Structured What-If Technique

One effective systematic method, a variation on which will be outlined further in this book, is the SWIFT or Structured What-If Technique. SWIFT is a systems-based risk identification technique that employs structured brainstorming, using pre-developed guidewords or headings in combination with prompts elicited from participants (which often begin with the phrases "What if…" or "How could…"), to examine risks and hazards at a systems or subsystems level [1]. The technique was originally developed as a simpler alternative to HAZOP (see Sect. 13.6.2).

The approach is flexible and has characteristics which happen to be conducive to managing risk in HIT systems:

- The involvement of multidisciplinary personnel and collaborative thinking
- A top down approach to focus on high level processes but without precluding the ability to drill down into detail
- The use of domain appropriate prompts or guidewords to illicit ideas for hazard register content

13.2.1 Using Guidewords for Creative Thinking

The cornerstone of SWIFT is the development of prompts or questions which are posed to stakeholders in order to precipitate impacts, hazards and causes. The most simple questions begin with 'What if…'. For example, one might ask:

- What if… the system became unavailable?
- What if… this information was recorded for the wrong patient?
- What if… this information was presented to the end user in a misleading manner?

These questions typically generate a torrent of useful responses when posed to appropriate experts and hazard register content can quickly be forthcoming. A key strength of this technique is that progress can be made very quickly and key areas of interest can be recognised early in the analysis for deeper study at a later time. Not all prompts require a what-if stem. Other useful questions can begin with 'How could…', 'Is it possible…' etc.

One shortcoming of the methodology is the ability to demonstrate completeness of the analysis. For example, how can one demonstrate that all appropriate 'what-if' questions have been posed? How does one know that the questions have been posed against all relevant parts of the system? There is no easy answer to this and ultimately the decision on completeness requires some judgement. Part of the solution involves the development of good quality prompts which, through experience, have been fruitful in eliciting useful material in the past. Secondly, it is helpful to be systematic about how the product is broken down into analysable chunks and to develop consistent angles of attack. A potential framework, but by no means the only methodology for doing this is set out in the forthcoming text.

SWIFT, as with many hazard analysis methods has its origins in industry and manufacturing. The guidewords originally conceived lend themselves to the management of raw materials, environmental release of chemicals, etc. In HIT the technique holds true but guidewords need to be modified. Card, Ward and Clarkson propose an alternative set of guidewords for SWIFT [1]:

- Wrong: Person or people
- Wrong: Place, location, site or environment
- Wrong: Thing or things (e.g. medication or equipment)
- Wrong: Idea, information, or understanding
- Wrong: Time, timing, or speed
- Wrong: Process (e.g. surgical procedure)
- Wrong: Amount

Note that Card's guidewords were primarily aimed at the analysis of general healthcare hazards rather than HIT but nevertheless seem a reasonable starting point. The list could be augmented with prompts taken from other classifications set out in this book. For example the framework proposed in Sects. 6.5 and 6.6 might produce guidewords such as:

- What if…part of all of the service failed
- What if…access was inappropriately denied
- What if…the identity of information failed to be preserved
- …etc.

13.2.2 The Multidisciplinary Approach

The content of the hazard register is often best developed as a multidisciplinary team activity and the simplicity of the SWIFT approach facilitates this. The technique is easy to learn even by those who are new to CRM and by applying it at a clinical and system business process level, the technique fits easily into a clinician's mind's eye of the working domain. The employment of a multidisciplinary team can also have an interesting side-effect. By engaging stakeholders in the analysis they are more likely to buy-in to the important task of risk management generally. With this comes an active interest in control implementation and a positive contribution to the development of a safety culture.

Clinical experts, product experts and technical professionals are all important contributors, each with their own perspective of potential hazards. Clearly there are practical limits on the numbers of contributors for an effective brainstorming session and a room full of several dozen experts is likely to yield slow progress. Small group sessions work well and, in some cases, it may be appropriate to form several workstreams for different aspects of the system under consideration.

Input materials are likely to include:

- an agenda which nominates a chair and minute taker
- a draft hazard register
- business process diagrams, use cases, functional descriptions, test plans, etc.
- an agreed risk matrix and risk management approach

Output materials might include:

- a more mature hazard register agreed and signed off by attendees
- an augmented project risk and issue log to escalate unmanaged concerns
- a work-off plan of outstanding activities
- a concise set of minutes setting out the decision that were made, where agreements were reached and the established next steps

When one begins composing the hazard register, initially there can be a barrage of divergent thinking, a whirlwind of ideas, scenarios and risks which can quickly forthcoming by those with domain and product expertise. A balance needs to be struck between a formalised, highly ordered approach and an unstructured free-for-all. Like any meeting, a strong chair-person with a solid understanding of SWIFT is essential to keep the session on track. A scribe or minute-taker should also be nominated (separate to the chair) and should consider maintaining notes in a manner that is visible to all attendees (for example by being projected onto a screen as they are written). Remember that demonstrating and documenting the fact that a systematic process has been followed is as important as actually following it. This develops the confidence we have in the final safety case. Good quality minutes from these sessions can be a powerful tool for justifying that confidence.

13.2.3 A Top-Down Methodology

The power of SWIFT rests with its ability to keep the analysis at a high level where this is needed. This abstraction might be criticised for lacking rigour and attention to detail but this does not necessarily follow. We need to ensure that the degree of analysis is commensurate with the level of risk to focus our efforts in the right place.

Thus the more quickly we can differentiate areas of greater risk, the sooner we can plan and prioritise our approach.

So we are left with the question, how can we establish which areas of the system have the greatest risk prior to undertaking the risk analysis itself? The trick is to focus attention on the resulting patient impact. By defining this we immediately draw a box around the likely upper limit of the risk in that area. One might, in many cases, determine that the potential safety impact of a system, module or function is relatively minor. Now the team could go on to spend many weeks and months investigating the failure modes for every transaction, operation, message and user interaction. One should deliberate on how fruitful this investment of effort is given the risk ceiling that has already been established.

The high level approach offered by SWIFT quickly weeds out those aspects of the system for which an in-depth analysis is frankly unwarranted. More to the point, it focuses the mind on those areas which would benefit from further analysis and, ultimately, the implementation of additional risk reduction measures.

In many analyses some areas of the system will be assessed as being associated with moderate or significant risk thereby justifying a deeper analysis. Interestingly one is able to continue to utilise SWIFT by simply applying it at a more granular level. For example, suppose an initial SWIFT analysis identifies that a particular module supports a critical clinical business process known to be associated with significant clinical risk. One might decide to list out all of the functions performed by that module and repeat the what-if. What if this function failed? What if it occurred against the wrong patient or at the wrong time? What if the output of the function was misleading?

Subsequently one might identify that a particular scenario contributes more risk than the others providing an opportunity to drill down even further. Ultimately one might decide to investigate individual database transactions, messages, error handlers or a component of system hardware using more focussed what-if scenarios to drive the analysis. With SWIFT a great deal can be learnt before we reach a level of deep technical detail. This is in contrast to many other systematic hazard identification techniques where practitioners can quickly get bogged down in detail from the start and fail to keep sight of the patient at the end of the harm storyboard.

Having said all that low-level systematic methods most definitely have their place in HIT CRM when applied appropriately. For example, a critical function which relies on complex messaging between two systems might justify the rigour and attention to detail afforded by a technique such as Fault Tree or Event Tree Analysis. Judicious use of these methodologies can yield results which would have been difficult or impossible to come by should SWIFT have been relied upon exclusively. Crucially, these techniques are far better able to demonstrate completeness of analysis and this evidence may be an important component of the argument set out in the safety case. In summary, SWIFT is a useful technique to balance effort and rigour whilst opening up the gateway to other methodologies where this is justified by the degree of clinical risk.

13.3 Breaking Down the Scope

Beginning a hazard register can be a daunting task and a place to start is needed. A methodical approach necessitates breaking down the target product into logical, bite-sized areas which can feasibly and realistically be assessed by a group of individuals. For large projects, this in itself can be challenging particularly as there are many different ways to slice the pie. The chosen categorisation might be by system module, function, supported clinical domain, etc. Whatever approach is taken each target should be clearly defined and ideally an analytical schedule constructed. For example, today we will analyse the requesting of radiology services, tomorrow scheduling radiology investigations, on Friday reporting radiology studies, etc.

One may choose to prioritise areas which are likely to be associated with the greatest clinical risk or, if necessary, those for which most information is currently available. In this way, the hazard register gradually emerges from a number of assessment workshops in a logical and methodical manner in association with the relevant experts.

Before blindly applying the what-if questions it can be helpful to consider further the different levels at which the analysis can occur. For example, suppose we begin by investigating the potential hazards in requesting radiology services. We might choose to apply our what-if questions to each system function and perhaps the user interface. But what about the hardware that supports the functionality, its configuration and messaging? Without a way to take into account all facets of the target application, it is easy to accidentally exclude areas that would have been useful to investigate.

Set out below are a number of analytical approaches which, taken collectively, can be used to tease out the hazards in a methodical manner. Each one essentially represents a different facet of the service, the alternative ways in which the same pie can be sliced.

13.3.1 The Clinical Business Process

Clinical business processes are the tasks that are undertaken to reach a particular care delivery objective. For example, booking an appointment, ordering an investigation, prescribing a drug and recording an admission all represent common processes with which most clinicians will be familiar. Each of these processes consists of a series of discrete steps involving a number of actors, one of which will typically be the patient.

For example in a prescribing and drug administration process a clinician might take the following steps:

1. Review current medications, clinical conditions and allergies
2. Determine clinical need for a new medication
3. Determine appropriate medication

4. Determine dose, route, form, etc.
5. Prescribe
6. Administer at appropriate time.

There may, in addition, be conditional branches in the process which only occur under certain circumstances, the medication is contraindicated, the patient is absent when the medication needs to be administered, etc. The processes can usually be mapped out quite easily using use case diagrams, flow charts, user stories or other tools. What is key when looking at the clinical business process is that the steps are agnostic about the tools which are used to achieve the objective, i.e. whether a particular electronic solution is utilised, paper systems or by some other means. The processes remain entirely in the clinical domain without any mention of screens, interfaces, data or hardware.

By knowing which clinical business processes a HIT system plans to support, the basic underpinning workflows can be ascertained. The workflows then offer a rich seam of potential candidates for hazard analysis. What is interesting here is the notion that it is possible to begin deriving hazard register content even before one considers a particular HIT solution. In fact, determining these high level hazards and impacts, divorced from the electronic system, has some significant advantages. For example, one will quickly develop a feel for the overall envelope of clinical risk. We can establish roughly where the system sits on the risk spectrum and therefore ascertain the approximate degree of rigour required in assuring the solution – all before we have a particular product in mind.

As the analysis is limited to the clinical domain, what cannot be established at this stage is a full set of evidenced controls for the system. It may of course be possible to suggest candidates for controls in the clinical domain (e.g. the availability of other sources of clinical information) but specific controls relating to the system itself would typically be no more than predictions or assumptions without a system (or its design) to analyse. Even in this case though, useful safety work can be conducted. One could for example formulate some characteristics of the system which, if implemented, could mitigate the causes and hazards identified in the clinical domain. These characteristics could then form the basis of safety requirements to be integrated into the general system requirements. In this way, a consideration of safety is included in the project from the outset influencing the design or procurement of a new HIT system.

13.3.2 The System Business Process

System business processes are the steps undertaken to achieve a particular objective in the HIT solution under examination. Thus, this aspect of the assessment allows the processes derived from the clinical business process analysis to be augmented with individual system interactions and behaviours. For example, consider the following flow:

1. Log into system
2. Search for and select the correct patient
3. Select 'medication – prescribe'
4. Search for the required drug
5. Select a drug from search results
6. …etc.

Note the key difference from our examination of the clinical business process; each step is a system interaction or system behaviour rather than a description of what individuals are doing or thinking.

The workflow should be broken down to manageable chunks each of which become a target for a round of hazard analysis and perhaps our what-if questions. It is at this stage that detailed hazards, causes and controls can be established which will form the bulk of the hazard register. The system business processes themselves may be derived from a number of different sources depending on the material available. Some systems may have detailed use cases with primary and exception flows carefully documented. For others the processes may need to be ascertained from training material, product descriptions or test cases.

Remember that not all system business processes are necessarily associated with an obvious clinical business process but they might still have the potential to impact clinical care. For example, there may be a system background task to find information in the database which has become detached from a patient record. Any association with a clinical business process is not immediately obvious but clearly, should the process fail to execute, this could contribute to harm (or at least represent an unreliable control). There is nearly always therefore the need to consider both clinical and system business processes in their own right.

Tracing the system and clinical business processes through in a logical end-to-end manner ensures that hazards and causes are not missed. The methodology itself is an important component of the SMS as its application demonstrates attention to completeness, logical thought and confidence.

13.3.3 The System Architecture

The system architecture represents the arrangement of physical and logical components that make up the HIT solution. These might include the hardware, networks, databases, storage components, peripherals and other infrastructure. Note that the system architecture should also take into account the interaction of individual software components and their interfaces which may also be a source of potential hazards and causes.

The need to analyse the system architecture is important. Not all potential failure modes are associated with business processes. The so-called 'non-functional' aspects of the system can represent a rich source of hazards and causes especially when looking at scenarios which relate to unavailability of the system. The architecture should be analysed to identify single points of failure and situations where one component is functionally dependent on another which itself may not be reliable.

Thought should be given to the system's redundancy and resiliency (see Chap. 7). Should a component fail, how soon could the service realistically be re-established? This analysis can be more complex when the system is hosted and/or managed by a third party organisation, in some cases reliability information might simply not be available. In these cases, organisations may need to implement 'blunt' controls such as the incorporation of contractual Service Level Agreements on the assumption that activities pursuant to these will mitigate the risk.

13.3.4 Interfaces and Messaging

The boundaries between systems and their components can create potential failure modes. This is particularly the case where the communicating systems are provided by different manufacturers and exchange safety critical information. In developing the hazard register, one should carefully identify the parts of the system which interface and consider the potential impact should the messaging between them fail. In some cases, the management of message failures will be the responsibility of a service management team. It is important to understand how these teams work, how will the failed messages be detected, will they be re-sent, will temporal order be maintained, etc.?

Perhaps the most important interface is the user interface with which the user operates the system. User interface design is a complex topic and beyond the scope of this book. However, it is key to understand that subtle features of the user interface have the potential to be causes of hazards. Is important clinical information difficult to access, truncated, inappropriately labelled or subject to excessive scrolling? Ideally one would identify these potential issues from design material or early prototypes. In some cases the full 'feel' of the user interface can only be assessed once the system has been at least partially built. Projects should therefore plan to assess, redesign and re-assess parts of the user interface where necessary and a CRM assessment of any shortcomings should influence the prioritisation of any areas for re-design.

13.3.5 Configuration and System Administrator Tasks

Not all system interactions are carried out by clinical users. Some individuals might be configured as super-users or system administrators and have access to functions which influence many aspects of the system. The functionality accessible to administrators needs to be included in the hazard analysis to determine whether it has the potential to directly or indirectly impact care. For example, a system administrator might undertake a monthly task to suspend system access for those users who have recently left the organisation. Should this task be undertaken incorrectly, the system administrator (or indeed the system) could incorrectly suspend one or more live user accounts. This could inadvertently deny them access to the service so appropriate checks and balances need to be in place to mitigate the risk.

Whilst system administrators often have access to high level functionality with potentially far-reaching consequences, the risk is often balanced against the fact that these users are likely to have had specific training and/or operational experience to develop the necessary skills. Nevertheless system administrator activity can easily be overlooked as a source of hazards if it is not specifically examined for each part of the system.

An additional level of complexity is introduced when one looks at the extent to which the service can be configured. HIT systems are becoming increasingly configurable and this degree of variability in different deployments of the same product can offer a rich source of hazards. During the initial roll-out of a product organisations to have to make configuration decisions which can extensively change system behaviour. These decisions may individually or in combination contribute new hazards or causes. One should therefore examine, for each part of the system, what configuration options are available and whether the decisions made for these settings could lead to harm. For large enterprise and national HIT solutions responsibility for configuration can cross several organisational boundaries. This can necessitate close co-operation between stakeholders especially where safety requirements differ.

It is always worth studying the potential implications of changing the configuration of the system during live service. Actions such as these should always be subject to formal change control and this is discussed further in Sect. 20.1. There is often merit in undertaking a proactive analysis of potential configurations would could introduce unforeseen hazards at a later date should they be implemented. For example, it may be recognised that a particular on/off configuration option would introduce significant risk if the setting were to be inadvertently changed at some point. By documenting this in the hazard register and safety case, this could provide a useful reference in the future when, maybe years after the safety case was originally authored, someone considers modifying the option.

13.4 Brainstorming Ideas for Hazard Register Content

Once the system has been carved into manageable slices the time has come to begin deriving material for the hazard register. Managing this critical task can be as much about good workshop management and crowd control as CRM. A common pattern which is seen at this kind of session is a period of unstructured divergent thinking followed by a period of convergent structuring and formalisation.

The divergent thinking phase can feel chaotic but should be encouraged, individuals should have a chance to 'get things off their chests' and produce ideas for later deliberation and analysis. The flurry of ideas for hazards, causes, controls and impacts need to be captured however disordered the thinking may seem. But this must be done on the understanding that, at some point, a methodical approach will be required in order to allow structuring of the hazard register and demonstrate completeness.

One good approach is to:

1. Generate and capture ideas
2. Structure ideas
3. Document
4. Revisit, analyse and complete

Let us consider using an example how we might illicit ideas for the hazard register and go on to structure them. Whilst the intention will always be to approach the task in a methodical way, in a group setting (especially with those not wholly experienced in CRM), ideas may be forthcoming in a more scatter-gun manner. The example here takes this into account and shows how a session leader might go about formulating a hazard register in the real world.

Suppose we were to begin to build a hazard register for part of an electronic prescribing and drug administration system. We might start by looking at the basic prescribing and administration business processes. Using the SWIFT (or a different) technique the assembled experts may generate the following information nuggets in the first few minutes of the session:

- A patient could be administered the wrong medication.
- A patient could be administered the correct medication but at an inappropriate dose.
- The user could select the wrong medication when writing the electronic prescription.
- The system could display the dose in a confusing way.
- A patient could develop medication toxicity if too much is administered.
- Staff administering a medication need to check they are giving it to the right patient.

All of the above statements are correct and indeed very relevant to the hazard register in question. However turning this into something that can be articulated as a well-constructed hazard register is not immediately obvious. Even at a most basic level determining which of these are hazard, causes, controls and impacts can be a challenge. So where to go from here?

Generally the easiest concepts to place are controls. Any characteristic of the system or its operation designed to reduce risk should be expressed as a control. In this case the need for staff to check they are administering medication to the correct patient is clearly a control, indeed a human factor control. This control will need a parent cause and/or hazard to belong to and, as yet, there are no obvious candidates. We should put this to one side for now.

Differentiating the other ideas as hazards, causes and impacts can be a little more challenging. Take the suggestion ‘A patient could be administered the wrong medication”. At first sight this appears to be a good candidate for a hazard, but one could argue that it is instead the impact of a hazard relating to the system displaying incorrect prescribing information. Alternatively could it be a cause of a hazard “Clinician makes prescribing error"? Very quickly one can become tied in logical knots and expend significant effort in structuring and re-structuring the argument.

One approach is to take the view that causes in some way relate to the HIT system or its operation, the impact relates to the patient's wellbeing and the hazard as the glue between the two. For example, with "Patient is administered incorrect medication" the focus is clearly on the patient and essentially divorced from any system behaviour which might give rise to this, it is therefore probably an impact. One might choose to expand its description to point out that the medication could have deleterious effects on the patient, might interact with other medication, a clinical condition or induce an allergic reaction. Similarly, failure to receive the intended medication could have adverse consequences. All the time, the ideas developed live firmly in the clinical domain and hence represent impacts.

Hazard	Impact	Causes
?	Patient is administered incorrect medication which could cause deleterious effects, trigger an allergy or worsen a clinical condition	?

There could be a whole host of system and human factor behaviours which could give rise to this impact. For example, the prescriber could select the wrong medication, a system defect could display a different medication to that which the prescriber chose, the drug dictionary or reference data could have become corrupt, etc. These scenarios all relate directly to the system or its operation and therefore they are good candidates for causes.

Hazard	Impact	Causes
?	Patient is administered incorrect medication which could cause deleterious effects, trigger an allergy or worsen a clinical condition	Prescriber selects wrong medication
		System displays wrong medication
		Local formulary data is corrupt

So where does this leave the hazard? The glue which binds all this together is perhaps the potential for the system to mislead someone into administering an incorrect or inappropriate medication. This simple description fits snugly between the identified causes and impacts. Importantly, the hazard description creates a tangible story within the medication administration business process which is easy to understand and contextualise.

Hazard	Impact	Causes
System misleads a user into administering an incorrect medication	Patient is administered incorrect medication whilst could cause deleterious effects, trigger an allergy or worsen a clinical condition	Prescriber selects wrong medication
		System displays wrong medication
		Local formulary data corrupt

Note that there is no suggestion here that any of this analysis is complete. There are likely to be many other causes and the group will need to return to this hazard to examine it in a more logical way. By structuring our early thoughts a picture of the hazard register for the product begins to emerge and acts as a seed for deeper and more organised examination.

Let's return to the original ideas the group proposed. We can perhaps see that there are at least a couple of other candidate hazards; one relating to the potential for the system to provide misleading drug dosage information and another whereby the system could fail to adequately identify the correct patient. Note that the latter hazard here relates to the control we previously put aside and has ended up spawning a hazard which will go on to have causes associated with it.

Hazard	Impact	Causes	Controls
The system could provide misleading dosage information	?	?	?
The patient to whom a medication is being administered could be incorrect	?	?	Staff administering a medication need to check they are giving it to the right patient

In this way a structure is starting to develop and the level of granularity between hazards is reasonably consistent. Importantly this simple structure would precipitate some new ideas; additional causes, more detailed impacts, etc. Although the process began with a divergence of ideas, structure and order has been resumed. Filling in the gaps now becomes easy and with further analysis completeness can start to be demonstrated.

One final but important point to note is that during brainstorming, hazards should not be dismissed purely on the basis that stakeholders believe that the risk has been mitigated. This rule holds true whether the mitigation is a design feature or a characteristic of the clinical environment. The argument often goes like this, "Well that would never happen because the nurse would spot there was a problem and take action accordingly…therefore that isn't a hazard". The logic is based on an assumption which may or may not be true, now or in the future. So long as a hazard is credible and in scope of the assessment it is has earned its rightful place in the hazard register.

13.5 Using Storytelling in Hazard Derivation

Prior to a hazard being fully described and assessed, it is essential to pin down exactly what the nature of the hazard is. This seems like an obvious thing to say but sometimes this can be a surprising challenge. During brainstorming a hazard may quickly be forthcoming but drill down into an individual's view on the nature of that hazard and you can soon discover that everyone sees it slightly differently. This difference might be subtle or sometimes gross but, in the early stages of hazard detection, there is likely to be some variability.

Resolving the nature of a hazard can give risk to much discussion but one tool appears to work well – the concept of a storyboard. When directors plan a movie scene or advertising commercial, their ideas often begin life as a series of comic-style drawings. Typically this will be a series of sequential boxes illustrating where people are standing, what they are doing, the kind of environment they are in and how the scene plays out. Each frame shows a progression in time with a beginning middle and end.

This construct can be surprisingly helpful when a group are trying to characterise and articulate a hazard. Of course, it's not usually necessary to employ an artist to illustrate this material. In practice, the frames can usually be described as word-pictures gradually building up the scenario. If an individual is struggling to communicate a potential hazard asking them to 'tell the story' starting with the triggering factors and ending with patient harm can often be highly fruitful.

So why does this technique work?

- Simplicity – The use of a storyboard is fundamentally simple and strips away technical complexities. Clinical personnel in particular are able to put themselves into the situation and consider how they would act under the circumstances. A story can be recounted by anyone and the listener can comprehend the scenario as the narrative plays out rather than having to construct it from disjointed detail.
- Context – As discussed earlier, a hazard is always a scenario rather than a thing, system, process or person. Using a storyboard forces one to provide context and describe the idea as a situation. For example, the starting point might be a confusing or badly rendered appointment booking screen in the application. It might be tempting to fall into the trap of deciding that it is the screen which is the hazard. By developing the storyboard one puts into context the way that screen is used, who uses it, under what circumstances, etc. The resulting, more 'rounded' hazard, perhaps describes the failure of a healthcare professional to correctly book an appointment – one cause of this hazard might then be the confusing manner in which dates are rendered on the screen in question.
- Human factors – A robust analysis is ultimately about how people deliver care and not necessarily just about how the tools behave. If you ask people to draw a business process they'll describe data flow, states and logical decision points. Ask someone to draw a storyboard and they show people, interactions and context. It's defining and capturing this latter information that makes for a good hazard description.

An important principle of the storyboard is that the final frame should show or describe the patient experiencing harm. If this frame cannot or is difficult to create this might suggest that the scenario is not actually a credible hazard. For example, an individual may suggest that a particular screen in the application is hazardous because it is slow to load – perhaps an intensely irritating behaviour for the individual reporting the issue. However, when this is put into a storyboard it might emerge that the particular screen is not used in a process associated with delivering care or indeed, that waiting a few extra seconds might be inconvenient but will ultimately not prevent the individual from gaining access to the functionality he/she needs. In this case, it might not be possible to describe a storyboard that clearly shows a path to harm. One might conclude that the scenario is in fact not a hazard at all but rather an issue of usability (which might still need to be addressed but for productivity rather than safety reasons).

13.6 Other Systematic Methods of Hazard Assessment

The SWIFT technique has been commonly applied to HIT in the UK and is taught on a number of training courses to support compliance with ISB 0129/0160 [2, 3]. Other techniques inherited from traditional safety-critical industries may be equally effective when used appropriately by experienced individuals. A selection are provided here to inform further adoption in healthcare.

13.6.1 Failure Mode and Effects Analysis (FMEA)

FMEA is a prospective hazard analysis technique which is widely used in many domains and increasingly in the service industries [4]. The methodology has its origins in military systems and the aerospace industry in the 1960s. Subsequently the automotive and chemical engineering sectors adopted the tool – indeed in some regulated industries application of the technique is now mandatory. The objective of the tool is to identify what in a product can fail, how it can fail, whether failure can be detected and the impact that will have. The technique can be supplemented with a Criticality Analysis which takes into account the severity of the failure. When this extension is employed, the technique is often called FMECA.

An FMEA team, usually put together for the purpose of the analysis, will work through a series of steps to apply the tool. At least one member of the team is usually an FMEA expert experienced in using the technique. A typical approach to an assessment might look like this:

1. Establish the known functions of the system
2. Assemble the multidisciplinary team
3. For each function or process establish the ways in failure might occur. Each of these items represents a failure mode (see Sect. 6.3).
4. Determine the effect of each failure mode
5. For FEMCA, ascertain the severity, likelihood and detectability of the failure mode and use this to generate the Risk Priority Number (RPN)
6. Rank the risks by RPN and use this as the basis for prioritising corrective actions
7. Implement the corrective actions focusing on changes in the product design and process improvement. Continue these activities until the risk is below an agreed acceptability threshold

Categories and scores are assigned to severity, likelihood and detectability measures (usually each has a scale of 1–10). The RPN is generated by numerically multiplying the three scores together. The inclusion of detectability in the formula is particularly helpful in reflecting the ability for users to mitigate risk in domains where this is practical (see Sect. 14.1.3). Without this one is forced to make detectability a contributing factor to likelihood or severity.

The technique has been adapted for use in a healthcare setting [5] (although not specific to the assessment of HIT). Healthcare FMEA or HFMEA has become an invaluable patient safety tool in many organisations with institutions such as the Department of Veterans Affairs (VA) implementing it in all of their medical centres [4]. VA has been able to demonstrate benefits in many diverse areas such as blood glucose monitoring, Emergency Department patient flows and MRI safety. Similarly the technique is advocated for use in England's Safer Patients Initiative and has been adopted by a number of hospitals [6].

There are a number of similarities between FME(C)A and SWIFT:

1. Each are preventative approaches ideally carried out during the design phase of a project
2. Both techniques benefit from collaborative effort involving multidisciplinary teams
3. Each has enough flexibility to allow analytical material to be updated as more is learnt about the system over time

An interesting point is whether the two techniques are alternatives or complimentary to each other. Potts et al. [7] undertook research to compare the SWIFT and FME(C)A techniques in a healthcare setting. A number of volunteers were asked to collectively undertake either a SWIFT or FMEA analysis on the same subject matter. Both groups provided positive feedback on the experience but obtained very different results. More than 50 % of the hazards identified were detected by one but not both techniques. In addition, the results varied significantly from the risks established by the research team using other systematic methods. The research suggests that use of a single technique alone may be significantly less effective than a combination of approaches.

FME(C)A, like all techniques, is not without its shortcomings. Benefits are dependent on the experience of the analyst, it is difficult to cover multiple failures and human errors and it can fail to produce a simple list of failure cases [8]. The technique focuses strongly on fault conditions and largely ignores scenarios where hazards occur during normal system operation with the service working as designed.

Criticism also arises when one examines the mathematics of the RPN [9] (which is also valid for a number of other techniques). Although in theory the RPN ranges from 1 to 1000, there are in fact only 120 possible scores. Similarly, failure modes which share the same RPN may in fact represent hazards of greater or lesser significance. These issues arise because the size of the gap between individual categories is undefined and, as such, multiplication makes little mathematical sense. This is not to say that this renders the technique useless but to claim that the analysis is in some way a proof is unjustified.

13.6.2 HAZOP

HAZOP is the acronym for HAZard and OPerability study. The method is a structured and systematic examination of a planned or existing product, process, procedure or system. It is a technique to identify risks to people, equipment, environment and/or organisational objectives. The technique was originally developed by ICI as a means of managing risk in industrial chemical plants.

HAZOP has much in common with SWIFT and FME(C)A, each of which employ an inductive reasoning technique. However HAZOP, as a consequence of its origins, lends itself to the analysis of process and in particular continuous processes. HAZOP also differs from FMEA in that the analytical team looks at unwanted outcomes and conditions and works back to possible causes and failure modes, whereas FMEA starts by identifying failure modes [10].

The input material for the analysis is variable and dependant on the nature of the project but procedural material is especially useful. The technique begins by taking apart a process and constructing a series of sections or sub-processes. The functional intention of each section is then agreed and documented. A team then subjects each section to a series of analytical guidewords and parameters to find potential deviations from the intention.

For example, in industry one might choose parameters such as flow, temperature, pressure and time along with guidewords such as too much, too little, none and in reverse. Each combination of parameter and guideword is then tested against the sub-process under consideration (e.g. low flow, high flow, low temperate, high temperature, etc.). HAZOP has been adapted specifically for assessing safety related computer systems where the technique is referred to as CHAZOP [11]. Additional work has been done to create more relevant parameters and guidewords to support its application to IT [12].

The use of HAZOP to assess healthcare and HIT risks is less well documented than with FMEA. This may reflect the technique's orientation towards the analysis of ordered and predictable processes which are characteristics that cannot be assumed in a healthcare setting. However, if one were able to construct a useful and relevant set of guidewords and parameters, careful application of the technique along with other systematic methods could prove fruitful.

13.6.3 Fault Tree Analysis and Event Tree Analysis

FTA is a top-down or deductive technique. It employs a graphical method to articulate the causes of a 'top event' which, in turn, is directly related to a hazard. Each cause of the top event is analysed further to examine its causes and so on. The resulting structure is represented as a tree, each event being linked by boolean logic operators (and, or, not, etc.). In some cases the top event can be replaced by a lower level event which effectively allows trees to be connected or embedded to model more complex systems. What emerges is a depiction of the system's resistance (or otherwise) to faults. Using the technique one is able to gauge whether there are any single points of failure and even model the effect of multiple failures (Fig. 13.1).

Fault trees are commonly used in safety critical industries such as aerospace. Their power is in being able to communicate complex failures in a simple graphical format which is relatively easy to learn. They can be applied to either potential failures or retrospectively in investigating actual failures. FTA has subtle limitations however especially when one needs to systematically identify all possible causes of a particular hazard – for this, an alternative technique needs to supplement the analysis. Fault trees are also notoriously difficult to apply to complex software.

Event Tree Analysis (ETA) is similar to FTA in that it uses a graphical representation of events – however in this case the analysis is essentially undertaken in reverse. The technique takes a starting event and looks forwards to examine what its consequences might be. The events depicted chronologically can include both fault and normal operating conditions. ETA doesn't require a hazard as a starting point and events can be modelled irrespective of whether or not they are related to safety. Assuming the necessary data is available, mathematical techniques can be employed to combine probabilities and derive an overall likelihood of particular outcomes occurring. ETA is particularly useful when it comes to modelling multiple layers of protection and where it is unclear at the outset what the end point might be.

Consider the example below depicting a tele-health solution remotely monitoring a patient's diabetes control (Fig. 13.2).

The use of FTA and ETA in HIT is limited by the fact that they essentially rely on components existing in a black or white state and the relationship between events being deterministic. So often in HIT one is dealing with shades of grey as a result of unpredictable user behaviour. In highly complex systems with many contributing

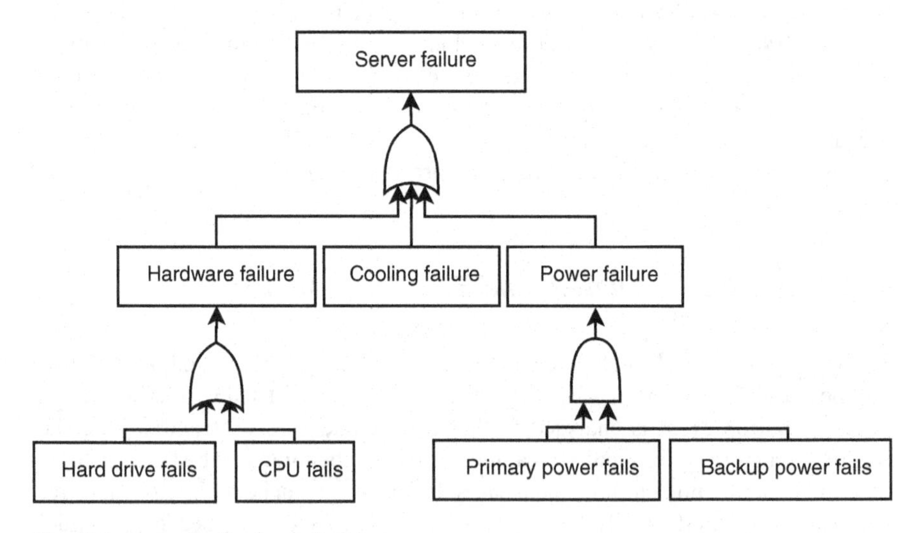

Fig. 13.1 Example of a simple fault tree

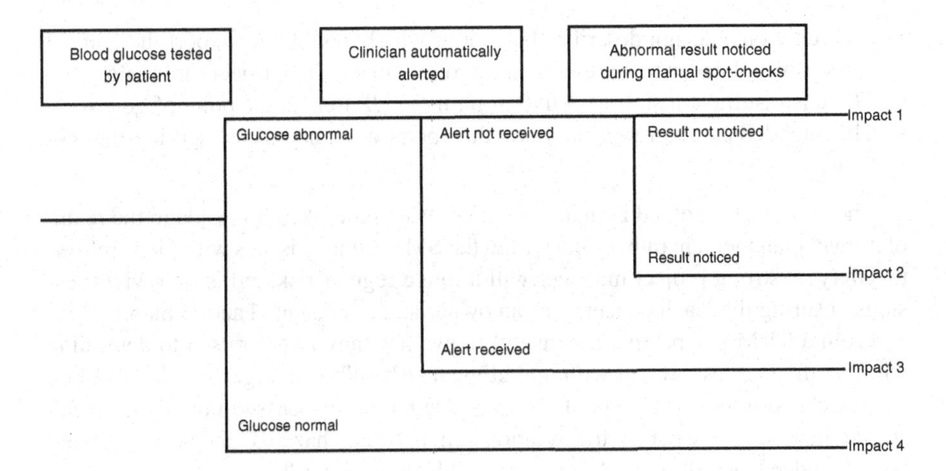

Fig. 13.2 Example event tree for a tele-health system monitoring a diabetic patient

components trees can sometimes become very large and complex and ultimately difficult to analyse and maintain.

Where FTA and ETA can be helpful is in assessing discrete parts of the system where failure is largely deterministic and the contributing causes straight-forward to define and model. For example, in analysing the failure of the application server, database, network or an entire data centre. FTA and ETA are techniques which can show great benefit when called upon in the right circumstances and where it is warranted by the degree of risk one is attempting to mitigate.

13.7 Safety-Related Issues

In the course of designing or implementing a HIT product it is almost inevitable that problems will come to light. In some ways the absence of any issues might suggest that the product is being insufficiently assured. Problems which threaten the project might be detected at any time during the design, build and test cycle and by any stakeholder. An issue is something which:

- Threatens the delivery, scope, cost, validity or viability of the project
- Was unforeseen or not fully understood until the project was underway
- Requires further management, escalation and regular review
- May or may not have the potential to change the safety position

For example, as an organisation learns more about a system in the course of a project it might be found that:

- The system architecture has a previously unidentified single point of failure
- Some client operating systems are of a version not supported by the application

- The functional design describes behaviour which would not support the clinical workflow and needs to be reworked (thereby threatening project timescales).
- There is insufficient staff to deliver training to all users at the point of go-live
- The intended configuration supports the operation of some care services but not others

The processes involved in managing these issues are likely to reside in the realm of project management rather than within the SMS as many issues will not be related to safety. A strong project manager will arrange regular risk and issue review sessions ensuring that each concern has an owner and documented action plan.

From a CRM perspective it is useful to evaluate any issues raised to determine whether they are associated with any clinical risk. What emerges is a list of open safety-related issues which should become the target of focussed investigation and risk reduction. So what is the relationship between hazards and safety-related issues? Indeed, are all issues hazards and all hazards issues?

If an issue could contribute to patient harm and it is unlikely to be imminently resolved, reflecting the problem in the hazard register can pay dividends to ensure that all stakeholder are aware and that the risk is formally assessed. One should consider:

- Does the issue represent a new hazard which had not previously been identified?
- Is the issue a new cause of an existing hazard?
- Does the issue make it more likely that an existing cause will be triggered to the extent that the risk changes?
- Is the issue actually best articulated as the failure of an existing control

Note that it is not always necessary to spell out the specific problem in precise detail in the hazard register. Good hazard registers have an appropriate level of granularity and often this level is above that of a specific issue. Thus what really matters is to ensure that a link can easily be drawn between an issue and some meaningful representation in the hazard register. Of course, as with any change to the hazard register, the degree of clinical risk may need to be re-evaluated in light of an issue.

The majority of hazards will not be issues and will not typically benefit from routine weekly review by the project team. Hazards can, in the main, be seen as entities sitting in the background characterising the risk in the event than an incident was to occur. Their purpose is to focus and prioritise the development of controls during the CRM analysis and to demonstrate the practical measures that have been put in place to reduce the clinical risk to ALARP. In contrast, issues are problems which require active management – they need someone to take ownership and run with the task of fixing them.

Occasionally one is faced with the scenario where the clinical risk associated with a hazard is unacceptable (i.e. where achievement of ALARP has failed). In this case it can be useful to promote the hazard to an issue. In this way the problem is ready to be picked up by the project team and be escalated accordingly.

Where an issue is complex, associated with significant clinical risk or needs to be communicated to a wider audience it can be beneficial to document and distribute a formal CRM assessment to project stakeholders. This is a convenient way of summarising the problem in one or two pages, setting out the rationale for the clinical risk evaluation and making any recommendations. The approach set out in Sect. 19.2 can be used for these purposes.

13.8 Summary

- To ensure that hazards can be reliably, consistently and completely identified, some kind of system is needed.
- There are a number of well-described, formal analytical techniques which can be adopted from other safety-critical industries. However some can be challenging to apply in HIT due to the extent to which human factors influence the risk.
- SWIFT is a useful technique to drive out hazard, causes, controls and impacts in HIT. It requires a multidisciplinary team and the application of guidewords usually in the form of 'what-if' scenarios.
- HIT systems are often large and complex and they need to be broken down into smaller chunks to enable a useful assessment to be carried out.
- Brainstorming and storytelling can be put to good effect in creating hazard register content. A strong leader is needed to guide contributors in the right direction and keep discussions on-track.
- Projects rarely run smoothly and issues often arise. Success requires these to be carefully handled by project managers and it is useful to examine whether any project issues have safety implications.

References

1. Card A, Ward J, Clarkson P. Beyond FMEA: the structured what-if technique (SWIFT). J Healthc Risk Manag. 2012;31(4):23–9.
2. Health and Social Care Information Centre. ISB0129. Clinical risk management: its application in the manufacture of health IT systems version 2. UK Information Standard Board for Health and Social Care. London. 2013.
3. Health and Social Care Information Centre. ISB0160. Clinical risk management: its application in the deployment and use of health IT. UK Information Standard Board for Health and Social Care. London, UK; 2013.
4. Stalhandske E, DeRosier J, Wilson R, Murphy J. Healthcare FMEA in the veterans health administration. Patient safety and quality healthcare. 2009.
5. DeRosier J, Stalhandske E, Bagian J, Nudell T. Using health care failure mode and effect analysis: the VA National Center for Patient Safety. Prospective risk analysis system. Jt Comm J Qual Improv. 2002;28(5):248–67.
6. The Health Foundation. Safer Patients Initiative. Lessons from the first major improvement programme addressing patient safety in the UK. London: The Health Foundation; 2011.

7. Potts H, Keen J, Denby T, et al. Towards a better understanding of delivering e-health systems: a systematic review using the meta-narrative method and two case studies. 2011. NHS National Institute for Health Research. London. http://www.nets.nihr.ac.uk/__data/assets/pdf_file/0003/81462/ES-08-1602-131.pdf.
8. Health and Safety Executive. Marine risk assessment. Prepared by Det Norske Veritas. London, UK; 2001.
9. Wheeler D. Quality Digest. Problems with risk priority numbers. [Online]. Cited 2015 July. Available from: http://www.qualitydigest.com/inside/quality-insider-column/problems-risk-priority-numbers.html, http://www.qualitydigest.com/inside/quality-insider-column/problems-risk-priority-numbers.html. 2011.
10. International Organization for Standardization. ISO 31010: Risk management – risk assessment techniques. Geneva; 2009.
11. Andow P. Guidance on HAZOP procesdures for computer-controlled plants. Contract research report No. 26/1991. Health and safety executive. KBC Process Technology Ltd. 1991.
12. Hulin B, Tschachtli R. Identifying software hazards with a modified CHAZOP. PESARO 2011: the first international conference on performance, safety and robustness in complex systems and applications. Munich, Germany; 2011.

Chapter 14
Estimating and Evaluating Clinical Risk

In previous chapters the focus has been on risk identification and the systematic methods used to characterise hazards. Our next task is to examine the practicalities of evaluating risk – studying the properties of hazards and their causes to establish the degree of risk and therefore its acceptability. Doing so allows us to prioritise those hazards which require further risk mitigation.

Note that some authors use the term 'estimation' for determining the degree of risk and reserve 'evaluation' for the process of determining whether or not the established risk is acceptable. In this text the term 'evaluation' will collectively refer to these activities combined.

14.1 Factors Which Influence the Degree of Clinical Risk

14.1.1 Clinical Dependency

Clinical dependency is a vitally important concept in evaluating clinical risk and should be a consideration in the assessment of all hazards. The term relates to the degree to which clinicians rely upon the system to deliver safe clinical care. The more the system is relied upon, the greater the impact on care should it become unavailable or deliver misleading information.

It is naïve to think that all HIT systems, in all care settings, influence clinical care to the same extent. The reality is that there is a broad spectrum ranging from those systems which hardly touch clinical business processes to those on which clinicians utterly rely. Similarly, the same system implemented in different settings may influence care to different degrees. Even when one studies individual users there will typically be variability in the significance that a system plays in their day-to-day delivery of care. Whilst one cannot separately evaluate a hazard for each individual

A. Stavert-Dobson, *Health Information Systems: Managing Clinical Risk*,
Health Informatics, DOI 10.1007/978-3-319-26612-1_14

Table 14.1 Relationship between role of HIT and clinical dependency

The system…	Clinical decisions are…
Co-exists with care delivery	Independent of system
Influences care delivery	Supported by system
Guides care delivery	Steered by system
Drives care delivery	Reliant on system

user, the variability should be taken into account in determining the typical or most likely impact of a hazard.

The degree of clinical dependency is strongly influenced by the role that the HIT system plays in a particular clinical business process. The role itself is affected by the functionality provided, the maturity of its implementation and the clinician's ability to access alternative sources of information. Table 14.1 outlines the relationship between a system's influence on care delivery and its ability to affect clinical decision making.

Note that it is neither right nor wrong for a system to be clinically relied upon. An assumption that a system is fundamentally unsafe if clinicians depend upon it is simply unfounded. A vital signs monitor may be relied upon exclusively to monitor the blood pressure of a post-operative patient but that fact does not make the device unsafe. What matters is to ensure that where a system is relied upon the clinical dependency can be justified given the risk. Often one finds that the higher the clinical dependency the higher the initial clinical risk. But, so long as appropriate controls are put in place to make the residual clinical risk as low as reasonably practicable then a case for the risk being justified can be made.

14.1.2 Workarounds

In deciding whether or not a system is relied upon it is interesting to think about the reasons why one wouldn't depend upon it based on the information it provides. For example, if a system or function is not relied upon to deliver care then either:

- the same information or function can be accessed in the system by some other means – i.e. a functional workaround can be put in place, or
- the information or function can easily be obtained in the clinical environment from a different source – i.e. a clinical workaround can be implemented, or
- the information or function it provides is simply not important to care delivery.

The first two reasons demonstrate the strong link between dependency and the presence of workarounds. Where one is able to bypass a defect we can, to some extent, mitigate the chances of it leading to harm.

Clinicians tend to be remarkably resourceful in their ability to continue care delivery in the absence of information and this should not be forgotten in the evaluation of risk and the development of the safety argument. Given a lack of information a clinician might telephone his colleagues, the patient's General

Practitioner, consult paper records or re-order an investigation. Falling back to these processes might be unpalatable when one has invested heavily in a HIT system however the risk analysis is not an exercise in evaluating project success but rather an assessment of the potential for harm.

But the real purpose of considering clinical workarounds is to tease out those scenarios where accessing an alternative source of information is simply not possible or realistic. The absence of practical workarounds can leave an identified hazard with a significant degree of unmitigated clinical risk. By documenting clinical workarounds as controls, it gives stakeholders an opportunity to challenge any assumptions or to drive the need for additional controls such as the maintenance of paper-based ordering processes, tandem data capture systems or service continuity policies.

Sometimes the system itself might still be capable of providing the information which is needed. It is not uncommon for parts of a system to fail whilst leaving other areas of the system functioning normally. For example, suppose that a system fails when it attempts to display a graph of clinical data. The system may also have functionality to view the underlying data in a tabular format. Whilst a numerical representation may be second-best in communicating a biological trend, the same information can still probably be gleaned with a little more analytical effort from the clinician. Whilst not ideal as a long term solution, some amount of operational redundancy reduces the clinical dependency on a single function. This, in turn, provides a degree of risk mitigation.

As well as the presence of a workaround, thought should be given to its real world viability. For example:

- How obvious is it to users that a workaround exists when they are engaged in a typical workflow?
- How inconvenient and time consuming is the workaround?
- How will additional cognitive load impact the user?
- How arduous is the process of communicating the presence of the workaround to users?
- Will the workaround cause any other deleterious effects on the system?
- Will diverting a user away from their normal workflow increase the possibility of a slip, lapse or mistake?

Thus there is a strong case for undertaking a risk assessment on the workaround itself to determine whether any introduced risk is greater than that being mitigated.

Note that whilst a workaround might not introduce clinical risk it could have other unwanted effects. For example, suppose a clinic which has been configured in the system suddenly becomes inexplicably unavailable for appointment booking. A system administrator might be able to functionally work around the issue by temporarily configuring a new clinic. However, it might be the case that this adversely impacts activity reporting and usage statistics with potential financial and capacity management consequences. In this case operational and financial risk have replaced clinical risk, a less than ideal situation which may not have been foreseen at the outset.

When it comes to finding alternative sources of information one should not forget that, in the majority of cases, there is a rich source of clinical data in the form of the patient. For millennia man has managed to deliver care in the absence of HIT systems by taking a sound clinical history from the patient and performing a thorough examination. As a risk evaluator one needs to step back and objectively review the overall clinical risk picture. Of course, in scenarios where access to the patient as an information source is not possible (e.g. language differences, unconscious or confused patients) relying on the patient as the sole source of information can be of limited value.

In all cases, the real-world clinical environment should be included in the equation when evaluating clinical risk – thus the importance of involving clinicians in the process. Clinical dependency should be assessed on clinical reality rather than an emotive assumption. For example, the management of the clinical emergency is sometimes cited as being critically dependent on HIT. But is this borne out in reality? Whilst there are occasions when a nugget of information can change the management of a critically ill patient, in the main clinicians will tend to get on and treat the patient in front of them on the basis of what they see. Unavailability of the electronic prescribing system for example will not prevent a clinician from administering adrenaline during a cardiac arrest, unavailability of the electronic protocol for cardiopulmonary resuscitation will not usually prevent cardiac defibrillation. In these critical situations, often the clinical dependency on the system is surprisingly low providing staff are appropriately trained.

The variation in clinical dependency and workaround presence over time is important as this will affect the assessment of future risk. Often when a system is first implemented, tandem sources of information continue to be available. Over time these sources may dry up as support for them discontinues. A typical example is seen when an electronic ordering system is implemented. To begin with the paper-based request forms are still available and continue to be printed and be replenished. In time clinicians develop trust in the new electronic system, no one complains when the paper forms run out and eventually staff forget that there ever was a form in the first place. When 1 day the system becomes unavailable and paper ordering is no longer supported, the impact is far greater than it would have been had it failed on day one of the implementation. It is an interesting paradox that those systems which are the most reliable are often associated with the greatest risk when they experience a fault.

By documenting controls such as the contingency use of paper forms organisations have a hook on which to hang supporting processes. A review of the hazard register may well point an individual to go and check if a business continuity measure referred to in the control still remains in place. Clinical dependency is therefore not an absolute entity that can be derived and then forgotten about but rather something that can change, evolve and require re-evaluation.

14.1.3 Detectability

Risk assessment is not a one-shot activity which we undertake blindly and never re-visit. As human beings we are constantly re-evaluating risk based on cues from our environment and modifying our behaviours accordingly. We instinctively make decisions about what information we should trust and what may be questionable. If a fault in our car satellite navigation system commands us to turn into a lake or a brick wall, most would (hopefully) question the validity of the advice. However, should that same system instruct us to turn into a road which turns out to be an unmarked dead-end we may well comply; at the time of making the turn, there is no evidence to suggest this is going to be unfruitful manoeuvre.

Research has shown that vigilance in detecting mistakes is highly effective in preventing incidents from turning into adverse events [1]. Clinicians will seamlessly fall back on their time-honoured clinical skills and professional judgement when information becomes unavailable or untrustworthy. However, to prompt this shift in mind-set, the clinician needs to detect that something is wrong. However resourceful an individual may be in finding a workaround to a problem unless he/she is able to perceive a fault, the workaround will never be executed.

This leads us to the notion of detectability – the extent to which we are able to identify deviation from the system's requirements or normal operation. Such is the importance of detectability that in some systematic methods of hazard identification such as FMECA detectability is included in the calculation of risk itself (see Sect. 13.6.1). In this way the lack of detectability is given an equal weight to likelihood and severity in deriving the Risk Priority Number, the main driver for prioritising corrective actions.

We discussed in Chap. 6 that safety-related issues are generally associated with either information which is misleading or information which is absent. To support the process of risk evaluation one can add a second axis to this classification, issues which are detectable and those which are not. The degree of clinical risk associated with a hazard will depend on its precise nature but, as a generalisation, the relative risk can be summarised as in (Table 14.2).

Issues which involve missing information but where that absence is obvious typically sit at the lower end of the risk scale. For example, when a system is completely unavailable this is clearly apparent and clinicians will automatically strive to obtain the information from elsewhere or delay making a clinical decision until the information is available. However, where information is unavailable but its absence cannot be detected, this will typically notch the clinical risk up a degree. For example, suppose an investigation result includes some important free text notes but these are omitted when reviewing the patient's clinical record. As not all results will contain free text notes there is nothing to prompt the clinician to call the lab, repeat the investigation or consult paper-based results.

A situation with a similar degree of relative risk exists when the information displayed by the system is misleading but in a way that is obvious. For example, suppose a trend line showing changes in plasma haemoglobin concentration over time on one occasion dips down to a value of zero (Day 21 in Fig. 14.1) whilst all other instances of the assay appear appropriate and correct.

Table 14.2 Relative risk of detectable and non-detectable faults

	Detectable issue	Non-detectable issue
Absent information	+	++
Misleading information	++	+++

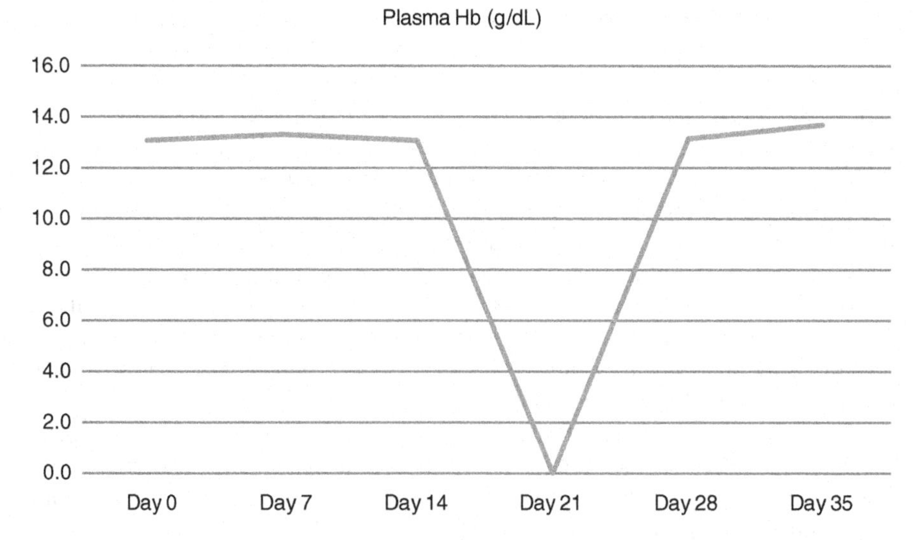

Fig. 14.1 Plasma haemoglobin over time showing an obviously suspicious result

Clearly something is wrong with the zero result. Whilst it might not be possible for the clinician to work out the nature of the fault, realistically he/she is unlikely to act upon the value or use it to influence a clinical decision.

The scenario becomes far more complex when one is faced with information which is both misleading and undetectable. Here there are no cues for the clinician to mistrust the data, no reason to consult other sources or investigate the issue further. The so-called 'credibly incorrect' scenario is often associated with the greatest degree of clinical risk as the human factor controls of professional judgement are effectively bypassed.

For example, suppose in the scenario above that instead of displaying an inappropriate value of zero for plasma haemoglobin the system reported an inaccurate value which happened to be just outside the normal range (Day 21 in Fig. 14.2). It is possible that a clinician would have acted upon that data at the time believing it to be accurate potentially resulting in harm.

An interesting observation with this kind of hazard is that it can be difficult for risk evaluators without domain knowledge to appreciate the full extent of the clinical risk. A call operator in the manufacturer's service centre is unlikely to be able to determine whether a particular example of misleading information could appear believable to a clinician. This is one instance when a clinical viewpoint is essential in prioritising a service incident.

In the design phase of the product lifecycle, it is useful to identify those data items which have the potential to significantly impact care should a credibly incorrect error occur. Assessors should ask themselves whether a data item is available elsewhere in the system to be cross-checked in the event that there is suspicion of inaccuracy. Similarly, are there single data items which, if misleading, could have a catastrophic clinical impact (e.g. a drug dosage or a positive indication of 'no

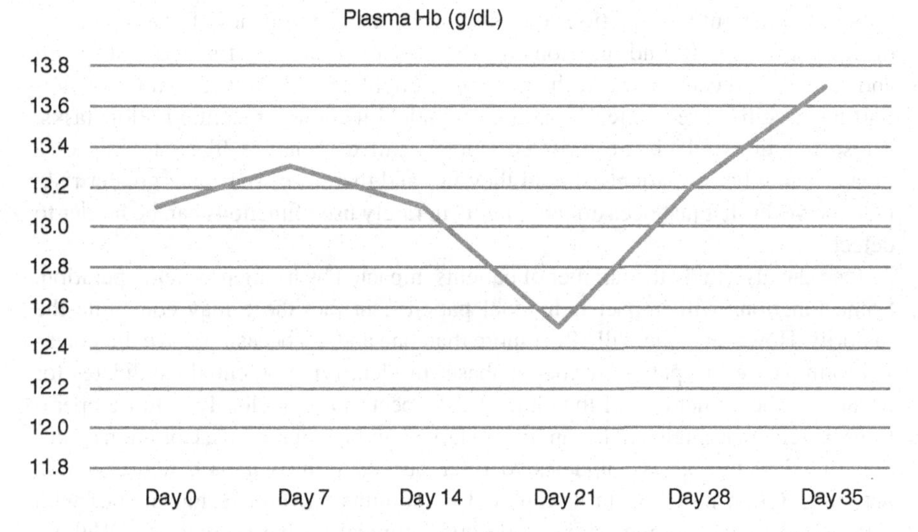

Fig. 14.2 Plasma haemoglobin over time showing an incorrect but perfectly credible result on Day 21

allergies' when this is not the case). These critical data items should be candidates for rigorous design analysis and testing. It might be appropriate for specific hazards to be developed to represent credibly incorrect failure modes in each of these areas.

14.1.4 Exposure and the Population at Risk

A more straight-forward factor in evaluating a hazard is an estimation of the number of patients potentially exposed to a particular issue, in other words, the scale of the problem. The greater the exposure to a potential fault or design flaw makes it more likely that this will trigger a hazard and thus the risk is raised.

Systems which are deployed nationally often contain the records of entire populations. Not only are databases large but the number of transactions each day may reach many thousands or even millions. In these goliath applications hazards which are considered incredible for individual transactions may be triggered several times a day when that likelihood is multiplied by real-world volumes. At the other end of the scale, small departmental clinical systems may only contain the records of a few hundred patients perhaps with a few dozen transactions per day. Whilst this somewhat over-simplified contributor to risk evaluation takes no account of the potential clinical impact of those systems, it still provides a starting point for establishing a high-level assessment.

Examination of the population at risk can also be applied at a more granular level. The functions of a system are not typically used (or at least relied upon) to the same extent across the application. Indeed should one be inclined, it might be possible to list out the functions of a system by usage frequency. Towards the top of this list one might find functions like logging on, searching for a patient record and reviewing recent notes. At the lower end might be found functionality for supporting specific rare clinical conditions, undo functions or configuration tasks. Those functions which are most commonly called upon are likely to impact a greater a number of patients should they fail and therefore contribute greater risk. Note however that just occasionally, faults in rarely used functions can be harder to detect.

One can also study the number of patients impacted by a single system operation. Some functions will impact individual patients; in fact these may constitute the majority. However some will affect more than one and, on occasion, have the potential to impact every patient in the database. In identifying potential candidates for hazards at the higher end of the clinical risk spectrum look closely at those operations which, on a single click, can impact large numbers of records. Common examples are data migrations, updates to reference data and database housekeeping activities. Often these are tasks initiated by administrative or technical staff with little or no understanding of the underlying clinical business processes. Without domain knowledge, individuals can fail to appreciate the risk and therefore the necessary degree of rigour in planning and executing these important operations.

User numbers also comes into play when assessing the effectiveness of human factor controls. Suppose a fault is identified and a workaround proposed which requires users to remember to do something specific. A control of this nature is likely to fail from time to time but how often? One factor to take into account is the number of users who will be required to execute the workaround. Firstly, communicating the advice to large numbers of users will be a significant challenge in most organisations not to mention how one might approach validating compliance. Secondly, where the workaround is confined to a small number of users it may be easier to integrate it into their daily workflow – especially where it is an operation which they revisit many times a day as part of their business function. In these circumstances it is easier to forge familiarity and compliance amongst small groups of users.

14.1.5 Complexity

In many ways complexity is the scourge of HIT systems and arguably all technologies. Health, even in comparison to other IT disciplines, demands the management of highly heterogeneous data. Clinicians are required to integrate laboratory results, the opinion of colleagues, captured images, testament of the patient and their family, surgical procedures, social circumstances, and so on. Medicine is a complex discipline in its own right so it is not entirely surprising that the HIT which supports its delivery can be equally complicated.

Whilst it is not inevitable that complex systems are associated with increased risk they almost always present new challenges both in their design and live operation. Where a complex system is also relied upon by healthcare professionals, a careful risk analysis must be undertaken to establish the potential effect of that complexity on the system, user and patient.

Definitions of complexity vary but a common theme is the inter-relationship of many, often disparate components with a degree of interdependency. Weaver [2] makes the case for two distinct types of complexity, organised and disorganised. Organised complexity arises from the non-random interaction of a system's parts. Whilst the behaviour of these systems may be challenging to forecast at the outset, ultimately modelling of the components in sufficient detail can allow us to observe and predict the resulting characteristics. In contrast, disorganised complexity is a characteristic of systems with large numbers of parts with seemingly random interactions. Analysis of these systems requires statistical and probabilistic techniques to predict overall behaviour.

In theory at least HIT systems should reside firmly in the domain of organised complexity. However, as systems grow and evolve the practicality of truly understanding a system's behaviour and make-up becomes an increasing challenge. Even small systems quickly reach a level of complexity where it becomes impossible for all aspects to be understood by any single individual especially when they are also subject to change. Analysis is then heavily dependent on the extent and quality of design documentation to be able to collectively predict behaviour in a particular set of circumstances.

The path from trigger to patient harm can be visualised as a series of events and conditions related through cause and effect. A safety analysis relies on being able to accurately model those relationships through inference. In complex systems it becomes increasingly difficult to visualise those associations and in some cases, even to predict those parts which influence behaviour at all. The socio-technical model (see Sect. 5.1) highlights the contribution of humans to the chain of events, entities whose behaviour is notoriously unpredictable. In these circumstances cause-effect relationships can quickly become blurred making it difficult to drive out the triggers of potential hazards. These complex pathways are said to be non-deterministic and can make it difficult or impossible to establish and characterise risk.

Where the potential impact of a hazard is clinically significant one might be required to overestimate the likelihood component of risk purely to accommodate the unknowns embedded within the complexity. System testing and defect resolution will of course go some way towards mitigating risk but again one is only able to test those scenarios which can be reasonably foreseen. In some cases one may need to rely on live service and operational experience to truly ascertain the risk associated with complex systems. This is one example where revisiting a risk evaluation over time incorporating lessons learned can pay dividends.

Complexity introduces additional challenges when it comes to evaluating faults and live service incidents. A key element in undertaking a root cause analysis is to be able to faithfully reproduce deviant behaviour. Faults in complex systems are more likely to be intermittent, unpredictable, non-deterministic and seemingly random. In particular it can be challenging to predict combinations of failures which might impact the system as a whole. Without an accurate set of pre-conditions on which to base the analysis any attempts to fix the issue will be severely hampered.

Even if a potential solution can be identified often the greatest challenge is in categorically proving that the fault has actually been fixed.

As well as the potential for systems to be technically complex, applications will vary in how difficult they are to operate by users. Where the user interface lacks usability operators will be increasingly dependent on their training to use it safely. A well-designed system will guide users into making correct operational decisions even if they stray beyond their existing operational experience and training. Whilst complexity does not by any means equate to poor design, a complex system is certainly more challenging for the user interface designer to get right. Employment of techniques such as user-centred design and iterative development come into their own in these scenarios.

14.1.6 Novelty

However rigorous a manufacturing and deployment project may be the reality is that doing things for the first time tends to be associated with increased risk. Novel systems have the potential to introduce risk in a number of different ways:

- Users are likely to be unfamiliar with new functionality, concepts or technology. This increases the likelihood of slips, lapses and mistakes and therefore risk. Whilst training is an important mitigation its effectiveness cannot compare with years of practical operational experience.
- One of the contributors to rigorous design is the incorporation of lessons learnt from previous projects and similar implementations. Unfortunately, unlike other safety critical industries, manufacturers of HIT systems tend to be reluctant to share their experiences across commercially competitive boundaries. Each system is often built from a standing start at the risk of duplicating faults already discovered and designed out by other manufacturers.
- The testing of novel systems can be challenging without a reliable benchmark for comparison. For example, when a system is upgraded it is usually possible to measure the performance of the new version and compare it with the old. With novel systems, how does one go about setting the level of performance acceptable to users in that domain? Indeed, which are the critical functions worthy of performance testing in the first place?
- Systems with a long track record are likely to have established controls in place for managing risk. Whether or not these controls have been formally documented in a risk assessment is in many ways less important. What matters is that over time a body of knowledge has been built up, the behaviour of the system has been characterised and the system's performance is predictable and deterministic. When this experience is unavailable to us, hazards can remain hidden or uncontrolled with an adverse impact on risk.
- Often one of the challenges of developing the safety case is in demonstrating that the proposed controls will be effective in the real world. Sometimes the only

way this can be done is by analysing the defects and incidents which are reported after go-live. Long established systems benefit from this operational experience and those controls which have been found to be ineffective are likely over time to have been replaced with ones which are. Novel systems do not have this history to fall back upon and the effectiveness of controls can often only be estimated.

It is worthy of note that novelty can present itself in many guises. In its most obvious form the service may be delivered using a brand new, disruptive, first of type technology. In contrast a healthcare organisation implementing an established technology for the first time will experience a different kind of novelty. Even rolling out an existing system into a new care setting or clinical specialty will come with a degree of unfamiliarity. Novelty exists on a spectrum but wherever it is encountered its impact on risk should be taken into account.

14.2 Bias in Risk Evaluation

An effective goal-based approach to risk management requires the application of judgement made by competent personnel taking into consideration the available evidence. But wherever human judgement is called for it leaves us susceptible to outside influences. Bias, in the context of CRM, is the tendency of an individual to make decisions based on factors other than objective evidence.

As risk practitioners we may fool ourselves into thinking that we are all capable of setting out a rational and logical safety argument uninfluenced by softer, less objective factors. And yet, often the first step to tackling potential bias is the recognition that our decisions are always susceptible to outside influences, either consciously or subconsciously.

Psychologists describe the notion of cognitive bias [3]. Cognitive bias is a limitation in our thinking which arises from our skewed perception of a situation, errors of memory or social influencers. Cognitive bias is thought to originate as an artefact of those strategies we employ to process information in emergency situations. Incorrect application or over-reliance on our innate response to threat can hamper objective decision making in circumstances where we should be taking account of the facts and evidence alone.

Interestingly humans tend to make inaccurate judgements in predictable ways. Psychologists have characterised dozens of different cognitive biases which humans tend to exhibit and some of these are relevant in risk management. For example:

- Anchoring effect – placing a disproportionate amount of weight on one piece of evidence or the first piece of evidence to be considered.
- Recency bias – Inappropriately favouring more recent events or observations over more historical data.
- Confirmation bias – Referencing information which supports our pre-existing views and neglecting that which does not.

- Normalcy bias – Refusing to believe that a set of circumstances could arise on the basis that it has never happened before.
- Observational selection bias – Believing that events are increasing in frequency when in effect we are simply noticing them more.
- Framing effect – Coming to a different decision depending on how the same information is presented.
- Risk compensation – taking greater risks as a result of the perception that safety has improved.

Commercial, delivery and political pressures are powerful drivers and from time to time we may be tempted to advise a course which is swayed by these contaminants. We may find ourselves being party to dialogs which include statements like:

- "They still haven't fixed this fault so let's raise it as a safety incident."
- "If we say it's a high clinical risk they'll have to fix it."
- "We can't say it's a safety issue or that will prevent the system going live."
- "If we admit to this being a safety issue now they'll wonder why we didn't spot it earlier."

Occasionally the carefully derived conclusions we draw from our safety analysis can be unpalatable to our peers, employers or customers and require a course of action which is costly or time consuming. From a personal perspective this can present a conflict when we are also faced with managing our careers, reputation and professional standing. It is human nature that sometimes it is easier to take the path of least resistance and derive conclusions which prove popular with stakeholders rather than challenge established wisdom.

To begin tackling biases we need to acknowledge that they exist. Once we accept this and begin to characterise their nature we can examine whether they are influencing our judgement. It can be useful therefore to map out these factors and determine whether they can be managed to reduce any potential conflict of interest. Who are the stakeholders who could influence me? To whom am I accountable? With what authority do I have to act?

For example, suppose you work for a commercial organisation producing HIT software. You occupy a number of roles within that organisation. As part of your annual performance assessment you have commercial targets to meet based on the number of organisations who successfully deploy your product. However, wearing a different hat, you are also responsible for carrying out a safety assessment of the product and identifying any safety related issues that could impact the go-live of one or more customers. It is probably only a matter of time before you encounter a situation where you have a conflict of interest. This completely foreseeable predicament could easily lead to (or be perceived to lead to) bias. In this case a simple reorganisation of responsibilities could be sufficient to eliminate the conflict.

Another tool in the management of bias is consultation and consensus. If an individual makes a judgement we disagree with it is easy to question that decision on the grounds of his or her interests or biases. By coming to a decision through a process of consultation, documentation and review an objective consensus can be sought. This is a particularly powerful process when those consulted possess a range of potential biases, ideally pulling in equal and opposite directions such that they are effectively cancelled out. In many cases consensus across organisational boundaries can provide the greatest confidence in the conclusions drawn.

That said, when it comes to bias and consensus a different evil can sometimes rear its ugly head – that of groupthink a concept first described in 1952 by William H. Whyte, Jr. in Fortune magazine. Groupthink or the 'band waggon effect' is a term used by psychologists to describe the situation where individuals' rational decision making process is hampered by their affiliation and conformance within the group. In other words a contributor may put aside their disagreements in order to maintain social cohesion – better to sit quiet and nod than to be seen as a troublemaker. The resulting lack of debate, challenge and dissenting viewpoint creates an artificial sense that the conclusions reached have taken into account the available facts and evidence. Groupthink has been cited as a significant contributor to a number of key financial and political decisions which in hindsight turned out to be erroneous [4].

Groupthink can be avoided by employing a number of strategies [5]. These include:

- Carefully managing the size of groups
- Using a facilitator to create a sense of openness and critical thinking
- Absenting of senior managers or those with a strong interest in a particular conclusion being reached
- The active examining of alternative points of view, devil's advocacy and the encouragement of challenge
- The consultation and inclusion of outside experts
- Avoiding beginning the analysis by setting out the conclusion one would like to reach

14.3 Summary

- In HIT there are a number of hazard characteristics which, if present, have the potential to raise the clinical risk.
- When clinicians rely on systems without access to alternative sources of information this can result in a single point of failure for the delivery of safe clinical care.
- Workarounds can help to mitigate risk but only when their implementation is realistic and a fault is detectable.
- Clinical risk evaluation involves a judgement call and this leaves the process susceptible to bias.
- Acknowledging external influences and the forms of bias to which humans are predisposed can help us to remain objective.

References

1. Magrabi F, Ong M, Runciman W, Coiera E. An analysis of computer-related patient safety incidents to inform the development of a classification. J Am Med Inform Assoc. 2010;17(6): 663–70.
2. Weaver W. Science and complexity. Am Sci. 1948;36:536–44.
3. Yudkowsky E. Cognitive biases potentially affecting judgment of global risks: machine intelligence research institute. New York: Oxford University Press; 2008.
4. Sorscher S. Group-think caused the market to fail. 2010. The Huffington Post. http://www.huffingtonpost.com/stan-sorscher/group-think-caused-the-ma_b_604810.html. [accessed 06.09.2010].
5. Hartwig R. Teams that thrive. [Online]; 2015 [cited 14 Apr 2015]. Available from: http://www.ryanhartwig.com/9-strategies-to-avoid-groupthink/.

Chapter 15
Developing Control Strategies

Controls are those measures which are put in place to reduce risk; arguably the most important elements of a hazard register. Although the structure of hazard registers vary, the objective should be to systematically mitigate each cause of the identified hazards. Some degree of traceability needs to be defined between the causes and controls to demonstrate completeness of this exercise. This assists in clearly establishing those causes for which controls do not exist.

In many projects the controls of most significance are those which traverse organisational boundaries. In other words where one party (often the software manufacturer) establishes that to mitigate the risk to acceptable levels another party (often the healthcare organisation) is required to put in place certain measures. These external controls provide an ideal starting point for downstream stakeholders to begin to build their assurance strategy on top of the safety work done by the manufacturer.

Controls exist in many forms and vary greatly in their nature – their effectiveness sits on a spectrum. Occasionally in the product lifecycle there is a need to implement measures to mitigate risk further. The process should begin with a control option analysis rather than simply plucking potential (and often the most convenient) controls out of the air. A sensible approach is to prioritise those mitigation strategies which are known to be the most effective and only move on to less effective options when implementation has been shown to be impractical compared the level of risk being mitigated. The aim should be to implement individual controls which are effective in their own right. Relying on multiple layers of ineffective controls is a poor strategy.

To this end, a classification system for controls and their relative effectiveness can be useful.

A. Stavert-Dobson, *Health Information Systems: Managing Clinical Risk*,
Health Informatics, DOI 10.1007/978-3-319-26612-1_15

15.1 Classification of Controls

There are a number of benefits in classifying controls. Firstly categorisation acts as a useful prompt to think about the full range of potential controls available. Of course, there is no requirement to have a control in each category but approaching control selection in this way helps in deriving mitigations that might otherwise have been overlooked using less systematic methods. Secondly, the methodology for validating controls may be common (or at least have similarities) across the class. This can greatly facilitate evidencing the implementation and effectiveness of controls.

Whilst there are many ways of broadly classifying controls, one system is commonly recognised across safety-critical industries. The system doesn't translate directly into HIT safety but with a little thought can provide a useful basis derived from existing good practise. At a high level controls can be categorised as follows [1]:

- Elimination
- Substitution
- Engineering
- Administrative
- Personal protective equipment

15.1.1 Elimination

Elimination is the most effective method of controlling risk, essentially reducing it to zero. To control a hazard in this way its causes need to be physically and totally removed. To take a non-HIT example, suppose one identifies a potential hazard in repairing an object which is positioned high off the ground. The hazard may be totally eliminated if the object being repaired was instead installed at ground-level. As one cannot fall off the ground essentially the hazard no longer exists and is therefore eliminated.

One might ask whether such parallels ever exist in HIT safety, after all we have already discussed the notion that risk cannot normally be eliminated in this domain. At a rather blunt level we may choose to control risk by making the decision to delay go-live of a HIT system and continue to operate with existing paper processes or legacy technologies. This approach may be unpalatable, expensive and impact the project reputation but ultimately this may be the only practical option available to us in some circumstances where the risk has been found to be unacceptable.

More practically we may be able to effect elimination by isolating just part of the system responsible for the risk. Most HIT systems are modular and at some level can be selectively switched off either through configuration or by the manufacturer updating the system code. One must examine however whether isolating the functional unit has the potential to introduce additional hazards by way of it being unavailable. An eliminative strategy is often only an option when the functionality in question is newly introduced. Removing some functionality from a new product might be acceptable to a customer whereas suddenly denying access to a well-entrenched component in an existing service could prove foolhardy.

One may also need to investigate whether the target functionality can really be truly isolated – advising users not to operate a particular screen does not represent elimination. A user could choose to ignore or fail to receive that advice in which case the hazard could still be realistically triggered. Thus elimination should involve a clearly-defined and well-engineered barrier which users are unable to circumvent.

Overall, elimination is a highly effective but often impractical means of control.

15.1.2 Substitution

In substitution the cause of a hazard is replaced by something which no longer represents a risk. In industry for example, this might mean substituting a dangerous chemical in a production process with a less hazardous alternative. In HIT we are sometimes provided with similar opportunities for control. For example, suppose a data entry form has a field for recording the name of a clinical procedure for a particular patient. The initial design uses a drop down list populated with pre-configured procedures. It soon becomes apparent that the number of possible procedures extends into the thousands which makes it difficult for the user to find and select the correct value in the list. A decision is made to mitigate the hazard by replacing the drop down list with a searchable field, aiding navigation and the likelihood of a correct selection. The original hazard is now strongly mitigated achieved by an effective substitution in the design.

Whilst substitution may not be as effective as elimination it is generally a sound alternative which preserves benefits realisation at the same time as managing risk.

15.1.3 Engineered Controls

In industry, engineered controls refer to those mitigations which are deliberately introduced to physically separate people from the hazard; for example, affixing a specifically designed safety guard to a hazardous piece of moving machinery. This definition needs some modification to make it useful and applicable in the domain of HIT. A reasonable definition might be those characteristics of a HIT system or its manufacturing process which pro-actively achieves risk reduction through carefully considered design. In other words we employ the principles of clinical risk management to foresee hazards and deliberately modify the requirements and/or design of the system to address those risks. In this way engineered controls tend to be properties of the system itself rather than the way it is operated or the environment in which it is implemented.

As with elimination and substitution, engineered controls are often highly effective. The difference is that engineered controls are less of a sledge-hammer – they are subtle, tactical and enhance rather than compromise the normal operation of the system. Engineered controls come in a number of flavours with varying levels of effectiveness and these are discussed in more detail in subsequent chapters.

15.1.4 Administrative Controls

Administrative controls are the measures we ourselves take to reduce risk in the course of our work and include the training we receive, policies, warning signs, etc. For example, suppose a clinician is reliant on a mobile device to gain access to clinical data – it is almost inevitable that from time to time the device will run out of power and access to clinical data could be compromised. One option could be to formally instruct the user during training to routinely place the device on charge every night.

The key difference between administrative and other types of control is the reliance on human operators to remember to act and behave in a particular manner. Unfortunately even well-meaning people are unreliable and forgetful particularly when other factors such as fatigue and extraneous distractions are in play. For this reason, administrative controls are considered weak and even multiple layers can be insufficient to mitigate significant hazards. When we find hazards which overly (or exclusively) rely on administrative controls, one should consider the true extent to which the risk is realistically mitigated.

In the example given above a better solution might be to provide easy access to charging stations, use a longer life battery, a lower power consumption device or have a supply of ready-charged emergency batteries available. Administrative controls should only be the option of choice when they complement more effective strategies or when all other realistic means of control have been exhausted.

15.1.5 Personal Protective Equipment?

The final category of control refers to the use of tools, apparatus and clothing which are designed to physically modify the human operator rather than the object he/she is working with, for example, the use of protective gloves or goggles. This concept appears to be completely irrelevant in the world of HIT safety as a class of control. However, taking a closer look there are some analogues at least at a metaphorical level if nothing else.

What we learn from protective equipment as a control strategy is the notion that there are fundamental properties of the operator which are important to risk mitigation. In fact every clinician carries with them the most important protective measures of all – knowledge and experience. In most HIT business processes the information flow nearly always includes a human being between the system and the patient. Irrespective of any information or lack of information provided by the system, ultimately the clinician's clinical judgement, training and professional experience provides a backstop mitigation to actual harm occurring.

This form of control in HIT can be considered as one which is omni-present and yet potentially fallible. Indeed, should a hazard be controlled exclusively by the application of clinical judgement one might consider whether the hazard is actually mitigated at all. Thought should also be given as to what degree clinical mitigation is possible and realistic in circumstances where the system provides misleading information which appears correct and credible. Unfortunately issues with poor detectability can often circumvent the application of clinical judgement.

Whilst clinical judgement provides risk control in practice thought should be given as to whether the liberal documentation of it as a specific control in the hazard register actually adds any useful learnings. Broad and non-specific controls such as this often introduce unnecessary noise rather than a tangible, evidential mitigation. Nevertheless clinical judgement is fundamental to the safe operation of HIT systems and its presence and application should at the very least be included as an assumption in the safety case.

15.2 Safety by Design

15.2.1 Active Engineered Controls

Active engineered controls are those features of a system which physically prevent a user from being able to take an action which could potentially be hazardous. Perhaps the simplest example is the use of validation on data input forms. For example, at the time of registering a patient for the first time a user will usually be prompted to enter the name and date of birth of the patient. A user doing this at speed could accidentally type the patient's name into the date of birth field. Many databases in the absence of being able to make sense of the entered data will record something like 1/1/1900 in the field by default – something which could subsequently be very misleading. A reasonable strategy is to introduce a design feature to validate the field in real time when the user moves to the next data input box or when he/she saves the completed form. The system might provide an error message directing the user to enter a valid date. A more advanced approach might be to validate not just the format of the entered data but determine whether it was in fact a viable calendar date e.g. April 31st or February 29th in a non-leap year.

Features such as these provide strong mitigation as the user is physically unable to bypass the feature. The downside is that these kinds of control require significant foresight, careful user-centred design and a high quality build. This often comes at financial cost and ultimately this needs to be balanced against the potential risk one is trying to mitigate. In addition one should remember that technical controls could introduce additional risk in their own right, both in normal and fault conditions. Suppose the validation is overly restrictive in our example? Perhaps it prevents the entry of a date of birth prior to 100 years ago which in an aging population could prove an inappropriate constraint. Perhaps a defect in the code always prevents the recording of 29th February even during a leap-year. Where there is complexity there is risk so the specification and testing of engineered controls should be at least as rigorous as for other safety-related parts of the system.

15.2.2 Passive Engineered Controls

Passive engineered controls are those characteristics of a system which attempt to forge safe operation through the provision of warnings, advice or decision support. Note that in this case the system doesn't go so far as to prevent a user from doing something but rather questions or brings to their attention an action which could potentially be hazardous. A simple example might be the soliciting of an acknowledgement, "Are you sure you wish to cancel all appointments for this patient?"

Examples of passive engineered controls are commonly seen in decision support. Suppose there is a hazard that a user could prescribe a drug to which a patient has a serious allergy. An active engineered control might automatically check the drug (and maybe the class of drug) against the patient's allergies and, after displaying an appropriate message, prevent the prescription being saved. In contrast a passive approach might involve the display of a prominent allergy warning indicator on the screen, the onus being on the user to observe this indicator before confirming the prescription. The functionality goes some way to addressing the risk but ultimately the user has to remember to make the check as part of their workflow.

Some passive controls will live outside the user interface and may not be apparent to day-to-day operators. For example, HIT systems typically need to exhibit resiliency in their architecture whether brought about through redundancy or other systematic means. These design features represent active engineered controls. However it is common for this to be supported by other more passive controls which require some degree of human intervention. The platforms on which systems reside can often be monitored for availability and performance. In some cases systems may be specifically instrumented to provide metrics on the execution of specific functions or the success of database transactions. Similarly systems may log errors or failed messages which are then made available for inspection by service management personnel.

These types of controls often span the gap between passive engineered and administrative controls. Whilst the system may be designed to report on unexpected behaviour, it requires human operators to provide the monitoring service and to remember to investigate system logs and queues. The users of course in this case are not clinical staff but rather highly-trained technical personnel responsible for maintaining system performance and often in the pursuit of meeting Service Level Agreements.

Unfortunately passive controls have a number of shortcomings perhaps the most significant being that they can simply be ignored and this fundamentally constrains their effectiveness. The more passive the control, the less likely it is to be attended to.

For example, a warning message displayed at log-on asking users to operate the system carefully is pointless as a practical control strategy. Effective passive controls are specific, triggered at the right point in the workflow under the right conditions and require at least some positive and audited acknowledgement from the user. Passive controls also need to be underpinned by training and local operating policies for them to be effective. Without this a culture of 'just press the okay button' can quickly develop if users fail to appreciate the reasons behind the need for the controls.

A second shortcoming is observed when passive controls are overused. It might be tempting in the pursuit of risk reduction to introduce passive engineered controls into all user interactions, perhaps a double or even a triple check each time a significant operation is carried out. In these circumstances users have a tendency to become alert-blind if they are prompted to confirm their actions or warned for each and every activity they undertake. It quickly becomes impossible to detect those prompts which are significant from those which represent background noise. Getting the right balance involves good quality user-centred design, a sound knowledge of the clinical business processes and an assessment of the potential risk associated with each system interaction.

Despite these shortcomings occasionally passive controls offer a more realistic mitigation strategy than active controls. There are times when enforcing a user action in the pursuit of safety is overly restrictive and just not pragmatic in the clinical environment. Often the freedom to over-ride a control in specific and agreed circumstances represents a control in itself. For example, suppose a system was to physically prevent the prescribing of a drug beyond a certain dose. There may well be circumstances when exceeding would what be considered the 'normal' dose range is clinically appropriate. In this case it would be more sensible to implement a passive control, asking the user to enter a reason for exceeding the limit rather than physically constraining their actions. What passive controls achieve to some extent is to share part of the risk with the user. In some ways the passive control is a vehicle for engaging clinical and technical judgement, providing the relevant individuals with a nudge to put into action their experience and knowledge.

15.3 Training and Human Factors as Controls

Most healthcare organisations stipulate how their HIT systems should be operated by users. Implementers needs to make decisions on how the various functions will be used to support care, what rules will be followed and what users are expected to do. This information then needs to be communicated in order to promote consistency and good practice from the moment of go-live. Occasionally specific human workarounds are needed to manage a defect, system limitation or implementation constraint. Indeed some behaviours will be critical for safety and it is these messages which benefit from careful communication. Healthcare organisations can reap the benefits of HIT systems earlier in the lifecycle when good quality training is provided. By integrating key safety themes into the training programme, users are armed with the knowledge they need to play their part in mitigating risk.

15.3.1 Identifying Training Controls

HIT system training programmes are often packed with information which can often challenge even the most adept learners. Not all information imparted during training is of equal importance and it can be difficult for leaners to differentiate those facts which are operationally critical from those which are advisory. If we accept that messages of safety and security should be high on the priority list then the course designer has a duty to develop creative strategies to promote retention of these facts. This is not an exercise in taking existing course materials and simply highlighting what the course facilitator might deem important. By the time a training programme is formulated, it is likely that an extensive hazard analysis will have already been undertaken. Many of the key human factor controls will have been painstakingly characterised and articulated in the hazard register. This is therefore the perfect place to start when it comes to teasing out those key facts which are critical for safe operation.

For each identified cause in the hazard register, one should be able to ask whether an element of user training is relevant to mitigating the risk. In some cases it will be appropriate to indicate specific controls to reflect this as a strategy. Note however that generic controls such as “User training required” are largely unhelpful as they contribute little to the argument and clutter the document. In contrast a more specific control might be documented as “Users should be made aware that when a patient has more than five recorded allergies, it will be necessary to click on the ellipsis ‘…’ to reveal the remaining details.” A control of this nature acts as an instruction to those building training materials to include and prioritise the instruction. Similarly those validating and evidencing training controls can easily create a test to check that the mitigation has actually been put in place.

Timing the delivery of a training programme is often a challenge. Train users too early and they deskill, leave it too late and coverage may not be achieved at the point of go-live. Adults learn best when the information presented to them is relevant and can be put to use immediately. Staff in different roles are likely to have a variety of training needs. Some users will require a significant depth of understanding, knowledge of the internal workings of the system and how it is configured. Others may only need knowledge of some basic functions to support their day-to-day work.

15.3.2 Classroom-Based Learning

Training users to operate a HIT system is only useful if the information can be retained and put to use in the workplace. Most healthcare organisations choose to implement some kind of face-to-face, session based training for mainstream HIT systems. Whilst this might be supplemented with other training modalities such as documentation, online help or e-learning, adults fundamentally learn best through experience and by reflecting on that experience with others. Imparting the key concepts of safe system operation should be a priority learning objective in any session-based learning programme. The more opportunity learners have to practise correct and relevant behaviours the more likely they are to retain information.

One of the more ineffective means of imparting key safety messages is the delivery of a solitary briefing as part of a chaotic staff induction day. Just occasionally a healthcare organisation takes the view that once a member of staff has been delivered information (in however a bland and disconnected manner) that it is simply the individual's responsibility to adhere to the advice henceforth. Any deviation from the instructed behaviour is taken to be a violation of the rules and is managed through disciplinary procedures. Such an authoritarian and threatening environment is not conducive to effective learning. Instead organisations need to understand their staff's training needs and provide the supportive surroundings which facilitate information retention.

Face-to-face contact with learners is resource intensive and potentially costly. It is therefore important that sessions are well-organised and timely with transparent learning outcomes. Leaders need to be well versed in the system under consideration, its safety features, hazards and controls. Sessions should be supplemented with appropriate course materials for learners to take away and use as reference in the future. Briefing users on safe operation of the system is usually most effective when it is integrated into general system training and undertaken in context with likeminded colleagues.

Adults absorb information differently from children and an understanding of learning psychology can be helpful when it comes to designing a training programme which ensures that key safety messages stick. Kolb describes a learning cycle for adults which consist of four discrete stages; Concrete experience, Reflective observation, Abstract conceptualisation and Active experimentation [2].

- Concrete experience usually takes the form of an active task which an individual or group is asked to undertake. This could for example be a scripted scenario which learners work through, operating the system from a set of click-by-click instructions. The learners get an opportunity to see how the system behaves under predictable and controlled conditions.
- Reflective observation is when we take time to think about what we have done, discuss that with others and learn from their experiences. In particular learners need to think about what was successful and unsuccessful. For example, a group might discuss how easy they found the scenario to work through, whether they needed to tackle any idiosyncrasies of the user interface or how the theoretical situation might be played out in the context of their real-world role.
- Abstract conceptualisation is when we use theory, background and knowledge to support what we have learnt and use these as the basis for new ideas. This might involve gaining a deeper understanding of what the system is doing behind the scenes, what functions are being triggered, what those functions do and what might be the intended and unintended consequences.
- Active experimentation occurs when we take what has been learnt and apply that to the real world. We might try out a new scenario of our own making it pertinent

to our role and determine whether the behaviour of the system under those circumstances confirms or refutes what we might have predicted.

Importantly individuals may have a preference for which of the four stages they prefer. Those who favour active experimentation will prefer to 'play' with the system, discover for themselves what the available functions do, make mistakes and learn from how the system responds. Abstract conceptualisers on the other hand will prefer to read the user manual, access on-line help and discuss the characteristics of the system from a more theoretical perspective. A training programme which facilitates a mixture of learning styles is likely to be more effective in gaining information retention across a group.

Those individuals leading sessions are more effective when they act in the capacity of facilitator rather than teacher or subject matter expert. This in particular should be borne in mind when it comes to imparting key safety principles of safe operation. Rather than restrict the discussion of system safety to a list of 'must do' directives, learners should be encouraged to think about potential hazards and controls for themselves. Facilitators can then reference back to the safety analysis and hazard register setting out the justification for the controls identified. In this way learners feel that safe operation is discovered rather than enforced and that the underlying rationale is owned rather than prescribed. Not only does this approach aid retention but learners are less likely to violate their training once they are party to the logic on which concepts are founded. This is particularly important when the actions required of users are contrary to existing practice, inconvenient or unintuitive.

At each stage of the learning process there is the opportunity for a facilitator to encourage discussion of key safety topics. During a concrete experience learners should be asked to think about what potential hazards might arise in those circumstances based on their previous experiences. Reflective observation provides individuals with an opportunity to think about the conditions which lead to a hazard being triggered, the experience of others in the group and how variations in the scenario might change the outcome. Abstract conceptualisation creates an opportunity to think about how data has changed since the hazard was triggered, what the impact of that might be for other users and how this could ultimately cause harm to the patient. Finally, active experimentation allows learners to think about different solutions and workarounds, their merits and shortcomings. Sessions which employ this method can be summarised as an experience for the learner rather than a lesson, a more fulfilling and ultimately memorable encounter.

Facilitators of course need to ensure that training does not become a grumbling session or indeed a hazard assessment workshop in its own right. Nevertheless it is not uncommon during training for those highly familiar with local business processes to identify new potential hazards or highlight limitations of the proposed controls. Facilitators should be given the opportunity to escalate these concerns to the CRM project team for potential assessment and inclusion in the hazard register.

In order to provide learners with practical experience of operating the system, some form of training environment is essential. Visualising system behaviour from screenshots or description alone fails to generate a sufficiently rich experience to promote information retention. Of course allowing users to freely experiment with a system which is operationally live can come with significant risks in its own right. But by simulating real-world experience one is able to create safety-related scenarios which allow learners to practise detecting hazards and manage those situations without impacting live care delivery. Users have the freedom to deliberately violate the rules and see for themselves how the consequences could adversely impact care. Indeed, simulating failure is commonplace in other safety critical industries such as aviation but is not widely used as a training modality in HIT.

Developing a system training environment comes with challenges. For example:

- The need for additional hardware, software licenses, configuration, testing, etc. all of which is likely to increase costs.
- The configuration of the training environment needs to reflect that of the live environment to provide learners with a representative experience. Prior to go-live often when the configuration is still in a state of flux this can introduce some complexities.
- Unfortunately HIT training environments often fail to include information that is in any way coherent or reminiscent of real clinical scenarios [3]. The training environment needs to be populated with data to the extent that it has the look and feel of a live system – this can be a time consuming task. For example, if the facilitator is pointing out the dangers of accidentally applying a filter to a list of data items, the list needs to be sufficiently populated to demonstrate the effect. Once a training course has completed there is usually the need to be able to reset the training data back to its initial state.
- Where applications have interfaces to other systems, simulating real-time messaging may prove troublesome. Of course if there is the possibility that live external systems could be contaminated with data from the training environment then this is a hazard in itself.
- It needs to be clear to users whether they are operating the live or training environment. If there is any possibility that these could be confused then this too represents a potential hazard.

Perhaps the greatest challenge for classroom-based training programmes is persuading staff to attend. However didactic a session may be, low training coverage will render important controls useless. This can be a particular challenge for clinical staff who not only find it difficult to secure time away from clinical duties but are then interrupted throughout. Compliance can be improved by:

- Scheduling sessions well in advance to allow alternative clinical cover to be put in place
- Promoting the benefits of the system which can be realised once training has been completed
- Communicating the importance of training in mitigating clinical risk
- Restricting access to the system until training has been completed

- Including the need to undertake HIT training in job descriptions, objectives and appraisal criteria
- Providing an enjoyable and useful training experience

15.3.3 Other Learning Modalities

Whilst face-to-face, session-based learning usually affords the best experience in terms of retention it is not always possible or practical to rely on this method of training alone. In particular users often only find gaps in their knowledge after training when they begin to use the system in anger. Having access to supplementary materials in the working environment is crucial if users are expected to solve safety critical problems themselves. Understanding a HIT system is a continuous process and users should be encouraged to revisit key concepts in order to validate their real-world behaviours.

Most modern systems are accompanied by user documentation of some description be it in printed or electronic format. Some may be supported by extensive and contextualised application help. In the clinical environment access to electronic materials is often more practical than trying to locate a long forgotten and out of date printed manual. In some cases, healthcare organisations may choose to create an online e-Learning package as an alternative or supplementary to face to face training. This approach can allow large numbers of users to be trained quickly and at a time when it is convenient for the learner. Where healthcare organisations choose to use these technologies exclusively to train users, the strategy should be evaluated against the extent to which human factor controls are relied upon to mitigate risk. E-Learning fails to immerse learner in discussion, concrete experience and experimentation which may adversely impact the retention of important information.

Careful thought needs to be put in to how key safety messages are communicated in training material. Often instruction manuals focus on the practical aspects of system functionality rather than the finer points of safe operation. This is especially the case when the only documentation available is that belonging to a manufacturer who undertakes little or no safety assurance. Sometimes the only reference to safety in the user guide is a disclaimer waiving the manufacturer's responsibilities.

Where specific behaviours are needed to control risk thought should be given as to how this is highlighted in documentation. Authors may choose different text styles, use of colour or boxed sections to draw attention to key messages. The language used should be appropriate to the audience with unnecessary abbreviations and slang terms being avoided. A long appendix of human factor controls disconnected from the main body of the document is likely to be less effective in supporting retention.

As discussed throughout this book, hazards often arise not from how the system is designed but from how it is configured and implemented. Controls relating to organisation-specific hazards are unlikely to be taken account of in the manufacturer's product documentation. Instructing on safe operation is best articulated when it is within the context of general operation but most healthcare organisations are unlikely to be able to penetrate the manufacturer's materials so this presents a challenge.

One solution is for healthcare organisations to create local Standard Operating Procedures (SOPs) which contain detailed, locally-tailored instructions taking system functionality, safety, local configuration and best practice into account. SOPs are the mainstay of many safety critical industries and are thought to significantly reduce risk from human error. SOPs are more than just a training manual, they allow for the processes set out within them to be formally governed and controlled. SOPs document the distilled wisdom of local experts, evidence and state of the art knowledge to bring consistency to the way that activities are carried out. They are formulated through careful consultation, analysis and agreement. New users have comfort that they have a single point of truth to refer to whilst experienced users gain from their colleagues acting in a predictable and consistent manner.

Providing that SOPs are accurate, implementable and adhered to, over time they become embedded within an organisation's culture and risk management is allowed to occur naturally and routinely. Of course creating this material requires a significant investment in time and effort to ensure that documentation is correct and complete. From there SOPs need to be distributed, communicated, implemented, monitored and maintained for them to remain effective.

SOPs however are not an alternative to the application of judgement and local process knowledge. It is impossible to create a SOP to deal with every possible contingency and variation of process. Users need to be encouraged to think independently in implementing SOPs rather than to follow them blindly with disregard for common sense.

15.4 Summary

- Controls come in many forms but differ greatly in their effectiveness.
- Controls are most effective when they are realised through careful design and engineering. To achieve this requires thought, domain knowledge, foresight and a culture of proactive risk management.
- Administrative controls depend on human factors and these cannot always be relied upon. They should only be considered when more effective controls have been ruled out.
- Most HIT systems require training for their safe operation. Training is an unreliable control but it can be more effective when the way users learn is taken into account.
- The hazard register is vital resource in determining the key messages that need to be put across to users during training.

References

1. Health and Safety Executive. Management of risk when planning work: the right priorities. Head office for HSE is Merseyside, UK. http://www.hse.gov.uk/construction/lwit/assets/downloads/hierarchy-risk-controls.pdf.
2. Kolb D. Experiential learning: experience as the source of learning and development. 1984. Prentice Hall. New Jersey. ISBN-10:0132952610.
3. Mohan V, Gold J. Collaborative intelligent case design model to facilitate simulated testing of clinical cognitive load. Portland: Oregon Health and Science University; 2014.

Chapter 16
Software Testing in Clinical Risk Management

16.1 Introduction

Testing is the process of determining whether a system meets it specified requirements. Put another way, we test a system in order to identify any gaps or errors in its specification or how it has been built. Standard ANSI/IEEE 1059 defines testing as "A process of analyzing a software item to detect the differences between existing and required conditions (that is defects/errors/bugs) and to evaluate the features of the software item" [1]. We test a system to check we have built something which is fit for purpose, is capable of supporting the anticipated number of users and, for existing products, that we have not broken something in the process.

Software testing is something which should be undertaken methodically and systematically. Simply asking the individual who developed to code to unilaterally test it as they see fit is likely to be problematic as a testing strategy. Testing is something which needs to be planned, executed and checked accompanied by documentation commensurate with the scale and complexity involved. The US Institute of Medicine has advocated that standardised testing procedures should be developed and used by manufacturers and health care organizations to assess the safe design, development, implementation, and deployment of HIT products [2].

The purpose of testing software is to perform verification and validation. These two concepts are different and understanding that difference is more than just a question of semantics. In order to assure a system and obtain the necessary evidence for the safety case, strategies for both verification and validation need to be in place.

Fundamentally verification asks the question "Are we building the product correctly". In other words if we accept that the requirements and specification are correct, are we creating a product which meets them? Is all the functionality present? Does the system behave in the way that was intended? Verification is an activity usually executed by software developers and generally doesn't require a deep understanding of the subject matter. Some level of verification can be achieved by examining documentation alone and doesn't necessarily require access to the system or its code.

A. Stavert-Dobson, *Health Information Systems: Managing Clinical Risk*,
Health Informatics, DOI 10.1007/978-3-319-26612-1_16

Validation on the other hand asks the question, “Is what we are building the right thing?” In other words, does the system actually meet the needs of the customer? Is it fit for purpose? Is it going to solve their problem? This requires taking a view of the system as a whole alongside knowledge of the domain in which the system is to be implemented. Validation requires judgement taking factors into account which are rarely binary.

To evidence the safety profile of a system both verification and validation are necessary. Verification ensures that the requirements which are critical for mitigating hazards are taken into account and successfully result in working code. It checks that those features which could introduce risk if not implemented correctly, function as intended. Validation provides a means of ascertaining whether design and engineering based controls are effective in the context of the wider system. It also provides a view on whether there are any hazards which were previously unforeseen and which might only become obvious once the big picture is clear.

Currently there are no specific standards which prescribe HIT software testing methodologies outside of formal medical device regulation. There are however a number of standards such as ISO/IEC 25010:2011 [3] which set out general requirements of software testing and quality management.

16.2 The Testing Process

16.2.1 Types of Software Testing

There are essentially two recognised types of software testing, static and dynamic. Static testing involves verifying the system (or parts of the system) without actually exercising it. This can take a number of different forms, for example:

- Document review and software walkthrough – where requirements and design material are methodically analysed for completeness and quality. Members of the development and wider project team are invited to suggest changes and enhancements to the material.
- Code Review and Inspection – where the software code is reviewed by developers to check the integrity, quality and logic in order to detect errors without actually executing it.
- Syntactic and logic checkers – tools to verify that the code being written makes sense, covers all eventualities and that the development environment will be able to compile the code when required.
- Compile checkers – tools which validate the code as it is compiled and report errors which must be addressed.
- Static code analysis tools – software which examines the code in detail and will identify poor programming practices and potentially ‘unsafe’ commands and procedures (for example the use of pointers or recursive functions).
- Formal verification – A series of techniques used to create mathematical proof that a piece of code will deliver predictable outcomes.

Dynamic testing occurs when the system code is executed and the software's behaviour analysed. This kind of testing can be done manually or automatically – often organisations choose to use both techniques to greater or lesser extent for a particular project.

Automated testing makes use of a software tool which is separate from the system under examination. The tool is configured to subject the target software to various inputs simulating user operation and then automatically compares the system's response to what would be expected. The advantage of this approach is that one is able to execute specific tests quickly and consistently with minimal user intervention. Hence this is an attractive method when faced with large systems with stable requirements in need of extensive and regular testing. Another feature is the ability to simulate many users operating the system concurrently without the need to employ large numbers of testers and associated hardware. An automated approach however has limitations when it comes to 'off-script' testing or in evaluating the more subjective characteristics of a system such as its usability and aesthetics.

Manual dynamic testing involves human operation of the system usually in a manner which is representative of the way users would interact with it in the real world. The tester subjects the system to a series of inputs, often directed by a pre-designed test script or user story and determines whether the system produces the response which would be expected based on the specification. In addition to formal test cases, testers are usually encouraged to use their ingenuity in not just necessarily following the user's expected workflow but also in performing unexpected or contradictory actions in order to try and 'break' the software.

The extent to which the tester has access to and is familiar with the underlying code gives rise to a further classification of testing. In 'black box' testing the assessor has little or no understanding of the internal workings of the system. The tester provides input into the application and its behaviour is observed and compared to the system's requirements, specification and design without any concern for how the software achieves its goal. Essentially this is a test carried out at the system's interfaces. This method has advantages in that those testing the software do not require an intimate knowledge of the technology or programming language. In fact knowledge of the subject matter and requirements may be of far greater value. Black box testing is useful in validating controls especially when they are articulated at a functional level. The main disadvantage is that without knowledge of the underlying code, it is difficult to guarantee that all code paths have been followed and verified. It can therefore be challenging to declare testing as being complete.

In so-called 'white box' testing, a different approach is taken. Here the tester has a sound knowledge of the system code, architecture and data models. He/she systematically develops tests to ensure that each piece of logic is validated. One popular technique is code coverage analysis whereby tools are employed to ensure that every code execution path is verified. Although this requires technical knowledge the outcome is much closer to a logical proof than a test purely against high level requirements. The techniques can be applied during unit testing as well as later in the testing cycle and in some safety-critical industries such as aviation, these forms of testing are mandatory. As formal logic is systematically employed, white box testing is useful in finding errors which could potentially lead to unforeseen circumstances and potentially hazardous scenarios. Of course one should bear in mind that a limitation of white box testing is that it is difficult or impossible to detect functionality which is missing from the product.

Manufacturers are ideally placed to undertake white-box testing as they have access to the system code and supporting materials. On the other hand most healthcare organisations implementing a HIT system will not have access to the detailed technical design and a white box approach might simply not be a realistic option in this case. Manufacturers implementing third-party components might also find themselves in this situation from time to time.

Finally there is a middle ground whereby the tester has some but not a complete knowledge of the systems workings. This is often referred to as 'grey-box' testing. For manufacturers the test strategy shouldn't be an either/or approach. Both black box and white box testing have their place and complement each other in achieving quality objectives.

16.2.2 Phases of Software Testing

Organisations tend to break software testing down into a series of discrete phases which subject the system to different kinds of examination. Often testing is conducted in parallel with build activities so that defects can be quickly identified and fed back into the development process. Each testing phase needs to have carefully planned entry and exit criteria to ensure that testing a phase is not executed on an immature build. In agile or iterative approaches, much of the testing is conducted within a sprint such that by the end of the short cycle a small but fully functioning system component is delivered. In more traditional waterfall methodologies the various phases of testing commence once much of the product has been built. Testing executed at any phase is a key safety assurance activity.

Software testing is frequently split into two elements, functional testing and non-functional testing. Functional testing looks at how the system behaves in response to user input. It takes into consideration workflow and the user's typical operation of the system. Functional testing provides evidence for managing those hazards related to misleading clinical data, usability, workflow and display. In contrast, non-functional testing looks at other characteristics of the system such as its performance under various conditions and how secure it is to malicious attack. This evidence supports mitigating hazards related to the system to responding slowly, denying access or becoming unavailable. Functional and non-functional testing needs to be carried out against the corresponding functional and non-functional requirement.

Functional Testing

Functional testing is itself subdivided into a number of phases. Each phase gradually increases in breadth beginning with verifying small units of system code and widening to validation of the overall system. The environment on which the system is running gradually needs to become more and more akin to the real-world platform on which the final version will sit.

Testing cycles often include the following phases:

Unit testing – examines a small piece of system code often executed in the development environment by the software developers. This is the most granular level of testing and scrutinises individual procedures, objects or classes. The intention is to verify the code at an early stage so that changes can be made quickly without impacting later test cycles. Unit testing guards against safety-related coding errors and verifies logic which could be integral in evidencing some technical controls.

Integration testing – looks at how the system behaves when individual units of code are brought together and allowed to interact. Essentially this is a test of the component interfaces to ensure that information is passed between them as intended.

System testing – considers how the system behaves as a whole and is usually done in an environment which is reflective of the live system. Whereas unit and integration testing is checked against the system's design and specification, system testing occurs against requirements specification. In that sense it is as much a test of the design and how the constituent components of the system have been integrated as it is the code and architecture. System testing is normally approached as a black box exercise where the testers are unfamiliar with the precise workings of the solution. Controls which are functional in nature are routinely validated during system testing.

Regression testing – checks whether functionality which has already been developed still works as expected when a change has been made. The change might take the form of an enhancement, bug fix or configuration update which runs the risk of breaking something which currently works. From a safety perspective regression testing checks that engineered controls are still valid and effective after a change has been made. As this kind of testing often needs to be executed regularly and consistently it particularly lends itself to an automated approach. Testers though need to remember the subtle effects of change on usability – this can be difficult to detect through conventional regression testing. A minor configuration enhancement may perform perfectly from a technical perspective but have disastrous consequences on workflow if the outcome is unplanned or the effect unpredicted.

Acceptance testing – generally involves the customer in making sure that what has been built is fit for purpose and meets their needs. Often the activity maps to a change in ownership or contractual milestone where the customer validates that go-live or further rollout can commence. This kind of testing requires expertise not in software development but in the domain subject matter and the business processes which the system supports. In the case of HIT it is often the clinicians and supporting staff of the healthcare organisation who are involved. Acceptance testing checks that the controls put in place are workable, do their job and meet the customer's safety expectations.

Non-functional Testing

Non-functional testing also takes a number of different forms. Performance testing looks at the speed at which the system operates and therefore how responsive it is to user operation and the execution of background tasks. Load testing examines performance when the system is subject to the maximum demand but within scope of the specification. Stress testing looks at how the system behaves in abnormal conditions for example if the specified load is exceeded, processors are removed or components deliberately taken off line. In practice, performance testing is usually carried out by employing specialist tools which simulate high demand.

16.2.3 The Test Environment and Test Data

As a rule, the environment on which tests are executed should be as close to the live operational environment as possible. This is particularly true in the latter stages of integration and system testing where a like-for-like comparison is needed. In practice obtaining such an environment comes with some challenges, for example:

- A test environment scaled to exactly the same specification as that supporting live operation is likely to be expensive.
- The live environment might include integration with other local or national systems which is difficult to simulate accurately especially under load.
- Local configuration of the application or the hardware on which it executes might not be replicable on the test bench.
- Peripheral devices used by a healthcare organisation might not be available for the manufacturer to integrate into the test environment.

In addition hazard registers often contain material describing the variability between different client devices, browser versions and operating systems. Evidencing controls in this area will require the software to be verified on each of these platforms to ensure compatibility. Systematically testing different combinations of platforms, configurations and devices can amount to a significant piece of work.

A challenge for most HIT systems is the difficulty in testing the system with data which reflects real-world clinical information. Often data inputted in the test lab is clean and complete – a far cry from the information which is typically captured in a busy clinical environment by users who are regularly distracted and may not be fully trained in the system's operations. This 'dirty' and often incomplete data may render very differently to the sterile test data deliberately crafted against the application's specification. Using extracts of real-world data for testing provides a more realistic environment but this can introduce confidentiality and information governance challenges and is only possible for technology which is already live.

The arguments set out in the safety case need to take account of the challenges faced in testing to ensure they don't undermine its claims. Any differences between the live and test environments will need to be acknowledged and assessed and one will need to justify that any dissimilarities will not change the testing outcome.

16.2.4 Testing Documentation

There is no right or wrong way to document the planning and execution of testing but a number of artefacts have become common over the years underpinned by the requirements in a number of different software testing standards. These artefacts are important inputs into the team developing the safety case.

- Test Strategy – Sets out the strategy and general approach to testing the system, the methods employed, roles and responsibilities, status reporting, standards followed, defect management plan, etc. In some cases this detail may be incorporated into the Test Plan.
- Test Plan – Summarises the planned phases of testing, testing scope, timelines, deliverables, pass/fail criteria and the test environments required. The test plan would typically be produced early in the development lifecycle around the time of requirements gathering and specification.
- Test Cases – Describes the scenarios which the system will be subjected to in order to validate and verify the functionality. Each test case takes part of a requirement and examines the behaviour of the system when a pre-determined sequence of inputs are executed. The tester then compares the expected outcome outlined in the test script against the actual outcome. Where the two match the test case passes, where there is a difference a defect is declared and recorded in a log. Increasingly test cases are formulated around 'user stories' which frame the test as a sequence of pretend (but realistic) clinical scenarios. This can help personnel to understand the context and subtleties of the domain.
- Test Report – Summarises the outcome of testing. In particular the report sets out how many tests were undertaken and the proportion which passed. The defects identified are usually outlined along with some indication of a fix priority. The report will normally declare any planned tests that were not able to be executed together with rationale or an explanation for this. The test report will conclude with an assessment of whether the test objectives set out in the test plan were completed and whether there are any outstanding activities.

- Traceability matrix – Provides logical linkage between the different components of testing. For example, traceability between the system's requirements and the test cases offers a means of evidencing completeness. When this is also linked to the defects found a rounded, logical picture of the system's validation can be demonstrated.

16.3 Testing as Evidence for the Safety Case

16.3.1 Test Strategy and Coverage

Early in the project a priority from a safety perspective is to ensure that the basic safety features of the system will be verified during testing. Without this assurance, the engineered controls described in the hazard register could not be evidenced as being either present or effective. Given that this type of control is typically at the stronger end of the effectiveness scale, failing to test is both a threat to the integrity of the safety case and a missed opportunity for valuable evidence. With that in mind, the hazard register therefore is a key starting point for formulating tests which specifically address important safety concerns. The intention, wherever practicable, is to show that the identified hazardous conditions cannot occur. Ideally one would ensure that it is impossible for the system to enter a hazardous state.

By tagging those controls which are safety-related one is able to derive a set of basic safety requirements. Safety requirements, when described in sufficient detail, are vital to testers tasked with formulating appropriate test cases. In this way solid traceability is created from the outset – simple but logical linkage between hazards, controls, requirements and test cases. Of course it is essential that the test cases created truly describe the expected system behaviour and that the integrity is not lost in traceability Chinese-whispers. It can therefore be useful for those formulating the hazard register to review the test cases and their validation criteria to ensure that the true nature of the test is correctly captured from a safety perspective.

To some extent this process can also work in reverse. Reviewing test cases helps us to think about how validating parts of the system might evidence controls in ways we had previously not considered. Similarly one might ask whether failure of the scenario contained in the test case might lead to harm should it occur in the real-world. This might lead us to formulate new hazards or new causes of existing hazards.

But testing has a role in safety which goes beyond simply validating specific safety requirements. The testing of complex systems often reveals defects that one would never have predicted at the design stage. These defects can only be teased out if all of the safety-related functionality is genuinely subject to testing. The scope and extent of testing set out in the test plan is therefore of great interest to safety case authors. A view needs to be taken on both the breadth and depth of testing. For example, does the scope of testing match the scope of the project and that of the safety analysis? Is the rigour employed in testing various components commensurate with the degree of risk assessed for those components?

By examining the test plan and test cases early in the development lifecycle plenty of opportunity is afforded to get the assurance strategy right and to formulate this from the ground up based on the safety objectives.

16.3.2 Artefact Inspection and Review

Project deliverables such as requirements, specifications and design material form the foundations of any system. Get these wrong and no amount of build quality assurance or testing will remedy the situation. The later a defect is identified, the more costly it is to address. There is therefore a strong business case and quality justification for investing considerable effort in ensuring that these artefacts are clear and complete at the outset. The process of inspection can go a long way to achieve this objective whilst simultaneously producing useful evidence for the safety case.

Inspection stems from the work of Michael Fagan at IBM in the 1970s. The technique is a rigorous form of peer review and involves a detailed examination of documentation either by individuals or groups of experts such as the security team, safety team or other subject matter experts. The intention is to detect potential faults, omissions or violations of quality and development standards. The activity seeks to forge a consensus amongst stakeholders to a point where artefacts are agreed and approved. Often the task is undertaken in a meeting environment with a nominated chairperson and minute taker. Defects observed in the documentation are noted for action and the resulting log may of course be referenced from the safety case as evidence of the activity.

A key target for inspection is the system's underlying requirements. Documenting sound requirements calls for a fine balance between providing enough detail for a designer to create the right product but without being overly prescriptive and dictate the means by which the requirement is fulfilled. Requirements for HIT systems arguably have a history of being vague and non-specific. The danger here is that the same requirement could be implemented in either a safe or unsafe manner. On inspecting documentation one should ask whether there is scope for a particular requirement to be misunderstood and implemented incorrectly. Is it ambiguous in any way? Does it allow for stakeholders to bear a different interpretation? In seeking clarity one can either reword the requirement accordingly, augment it with further clarification or trace it to specific safety requirements which educate designers on potential hazards.

In a complex environment like healthcare delivery, often the key to good requirements is the provision of context and background. For example, suppose functionality is required to book an urgent appointment for a patient. One could leave the requirement at a high level but more information and context will likely result in an improved design. Perhaps urgent appointments occasionally need to be booked in an environment where actual contact with the patient is just not practical. Maybe some users need to reliably perform this operation in a mobile or remote setting. In many ways deciding how these factors might change the design is less important than providing the necessary information to the designers to help them to make optimal decisions. Inspection of the requirements by subject matter experts should reveal material which lacks this detail or is ambiguous and therefore needs further explanation and context.

Inspections need not be limited to the system's requirements, in fact any artefact which is human-readable can be subject to analysis. For example:

- Design material
- Test plans
- Test cases
- User documentation
- Standard Operating Procedures
- Configuration specification
- Transition strategy
- Decision support rules

One final candidate for inspection emerges in the complex area of validating a data migration between two HIT systems. When data is moved from one system to another there is significant potential for it to be transformed incorrectly or be incomplete and this can be associated with clinical risk (see Sect. 20.3). Even generating rules by which the data can be validated can be a challenge when one considers all the possible variations. Part of the solution might be to simply provide the dataset to expert users for them to inspect – this could be in the form of a spreadsheet or database table to assist filtering and sorting. Administrative users often become very familiar with their local data and can quickly spot erroneous fields or defects which would otherwise be difficult to detect. It can also be a useful exercise to gain a view on the quality of the underlying data. Where defects are found it may be possible to enhance the data transformation rules or prompt updates to the dataset in the source system. Of course the practicalities of this approach will vary depending on the volume and complexity of the data.

16.3.3 Review of Design Decisions

In constructing and configuring the system it is likely that key design decisions will need to be made by both the manufacturer and healthcare organisation. Evidence that those decisions have been made in a considered and informed manner through consultation with appropriate experts lends weight to the safety case. In some cases, the manufacturer may make design or configuration recommendations based on their own risk analysis or operational experience. Healthcare organisations need to ensure that where they deviate from these recommendations that any potential impact on clinical risk is discussed with the manufacturer and factored in to the evaluation. In particular the system should be tested in an environment where those design decisions have been implemented in order to provide a holistic picture of the final solution.

16.3.4 Management of Defects

Testing of course does little or nothing by itself to mitigate risk. Only when coupled with effective and timely defect resolution is progress made. Finding defects during the testing process is inevitable – in some cases one could argue that failing to detect defects probably means that the system has not been tested with sufficient rigour rather than being any testament to the quality of its build. Defects are of great importance to safety case authors. Much proactive safety assurance work is concerned with what might happen if the system behaves unexpectedly. But when faced with a known defect, the path to harm is significantly shortened. For this reason defects warrant specific examination in their own right.

There are two primary purposes in undertaking such an assessment. Firstly the period between test and go-live can involve an iterative cycle of defect identification, fixing and re-testing. In many cases it is simply not possible to address all defects in time for go-live and organisations must prioritise fixing those issues which present the greatest threat to the system's success. Prioritisation should be more than addressing only those issues which threaten to stop the software's execution. The degree of clinical risk introduced by the defect should always be a key factor in evaluating its significance.

Secondly, it is important to know the potential clinical impact of a defect and its associated risk should there come a point where a case is made to go live with the system without the defect being fixed. This critical point of a project is not the time to begin a safety assessment of the issue. Instead an objective evaluation of a safety-related defect should be made early such that the starting point for the argument to go live is made from a position of knowledge fully cognisant of the degree of clinical risk involved. To achieve these objectives a defect should be assessed on the assumption that it will not be fixed prior to go-live. Certainly it is overly confusing to contaminate the 'likelihood' component of the risk calculation with the probability of the issue being fixed in time.

Whilst there are benefits in safety assessing defects early, if one undertakes the task too eagerly this can result in impossible mountain of analysis. Early in the testing cycle, for example in unit testing, defects will be raised and fixed on an almost daily basis. Attempting any kind of meaningful safety assessment in at this stage is largely a waste of resource. Plan instead to assess the outstanding test issues at a point where there are a realistic number of issues (given the resource available to assess them) but still plenty of time to fix those defects which introduce significant clinical risk.

A traceability matrix comes into its own when managing defects. If, at the planning stage, requirements, test cases and hazards have been mapped then it is a relatively straight-forward task to link the defect back to the hazard and understand the potential impact on the patient. The mapping also makes it easier to clarify the safety requirements which must be taken into account when testing the rectification. For large and complex systems in particular or those with aggressive, iterative life-cycles, investing in solid traceability can pay substantial dividends.

The techniques for carrying out a safety assessment on a defect are largely the same as for evaluating the risk for any hazard or indeed an incident in live service. At the very least one should consider:

- Whether the defect affects a clinical business process known to be associated with significant clinical risk.
- Whether a clinician would rely upon the part of the system affected to deliver safe clinical care.
- The frequency with which the defect would be triggered during normal operation of the system.
- The scale of the issue in term of how many users and/or patients could be impacted.
- Whether the defect could realistically be worked around by users.
- Whether the problem would constrain new functionality or would break existing functionality on which users are critically dependent.
- The extent to which the problem could be difficult to detect by a user and therefore fail to prompt them to find alternative solutions.
- Whether the issue could result in a situation where the information presented causes a clinician to make decisions based on misleading data.
- Whether working around the defect would be reliant on human factors which might bring into question the likely reliability of the workaround.

Organisations should decide in their SMS how defects are assessed and documented, preferably using a common format or template to forge consistency. The goal should be to communicate to stakeholders an unambiguous view on the clinical risk and whether that risk is likely to change over time.

Healthcare organisations have a responsibility to understand what defects currently exist in the product and asking the manufacturer's view on the clinical risk associated with known issues is a reasonable request. Defects found by the healthcare organisation could either be specific to their configuration or more general product issues. A detailed safety analysis is particularly critical when the manufacturer is involved in providing a fix. Manufacturers are not always aware of the clinical implications of a particular defect. A dispassionate explanation of the hazard will help them not only to prioritise it appropriately but also to suggest potential workarounds or clarify any functional misunderstandings.

The occurrence of defects opens up interesting insights into precisely how the system can fail and what the impact on other components of the system might be. It is quite possible that these failure modes had previously not been imagined or considered credible. In these circumstances it is often appropriate to update the hazard register to reflect these learnings. Even if the issue is fixed the potential for unexpected failure in a safety critical area deserves acknowledgement in the hazard register to preserve these insights for the future.

16.3.5 Documenting Test Activities in the Safety Case

In summarising testing there are largely two objectives:

1. Demonstrating that significant rigour has been applied to give confidence that the major defects have been found and managed.
2. Setting out the rationale and justification for the residual clinical risk associated with any outstanding defects.

In achieving the first goal a summary should be provided in the safety case on the general approach to testing. Whilst re-iterating the entire test plan is unnecessary, summarising the key safety-related testing activities in a logical manner builds confidence in the safety case's claims. In particular the report should show that the breadth of testing covers all the safety-related components and that the depth of testing in key areas is commensurate with the risk assessed. Evidencing completeness can be achieved by outlining the means by which test traceability from requirements to test cases was achieved and a reference to the completed traceability matrix would be welcomed by any auditor.

Where in the course of the project the CRM team has noted deficiencies in the test strategy and instigated corrective actions these interventions should not be hidden away but rather brought out in the safety case. A reader will interpret this quite correctly as the application of an honest and diligent process rather than any confession of wrong-doing or slack-handedness.

A flavour for testing rigour can be communicated by indicating the number of tests executed, especially when this is compared to the number of outstanding defects. For example, showing that 1,000 tests were completed and 10 failed shows a very different picture to circumstances where 15 tests were completed and 10 failed. Where it was not possible to run a test for some reason this should be explicitly recorded. Any decision not to execute a test needs to be underpinned by sound justification and without compromising any evidence on which the safety case may rely.

Summarising any outstanding defects at the end of testing is a key component of the safety case – indeed it cannot be considered complete without this key data. This section of the report should set out the remaining defects along with the hazards affected and degree of clinical risk involved in continuing to go-live. For significant risks, the safety case should detail the precise nature of the defect, any controls or workarounds which stakeholders may need to put in place and recommendations for how the issue should be managed. Section 19.2 describes a framework for documenting issues in the live environment and a similar format can be harnessed for outlining residual defects in the safety case. Where discussions are held to ascertain whether it is appropriate to progress with go-live, these sessions should be minuted so that the conclusions can be referenced and summarised in the safety case.

An example of a basic defect summary is provided below:

Requirement 148 specifies that the system should display a graph showing the trend in a patient's recorded blood pressure over time. A defect has been found in this functionality.

Clinicians need to be able to view both the systolic and diastolic blood pressure values on the same graph concurrently. During user acceptance testing it was identified that the user could view one or the other parameter but not both at the same time. This functionality will be used by clinicians in approximately 50 % of consultations. The defect could compromise a clinician's ability to correctly interpret changes in blood pressure which could result in the delivery of incorrect clinical care. The issue would be easily detected by a trained user.

It should be noted that alternative functionality exists to display a tabular, numerical representation of the data where both systolic and diastolic values are correctly displayed side-by-side. Furthermore, this screen is encountered in the workflow immediately before the affected graphical representation. Users could work around the issue by using the tabular view until the issue is fixed. However access to the graphical view cannot be technically prevented.

The clinical risk associated with this issue has been assessed as Severity: Moderate, Likelihood of harm: Remote, Clinical Risk: Low.

The manufacturer will deploy a fix for this issue 4 weeks after go-live. It was agreed at the project board on 14/03/14 that the system can continue to go-live as planned (Reference: PBMinutes140314). It is recommended that all users are made aware of the limitations of the system in this area and that they should use the tabular view of blood pressures until the defect is fixed. The hazard register has been updated accordingly.

16.4 Summary

- Software testing is a common control in many hazard registers but for the safety argument to be valid, this must be backed up by solid test evidence.
- The test strategy and test cases should be informed by the safety assessment to ensure that focus is placed on those areas know to be associated with the greatest risk.
- Testing can be more than manual operation of the system under simulated conditions. Inspection of key artefacts such as requirements, designs and specifications by suitable experts is a widely used technique in other safety critical industries and offers valuable evidence for the safety case.
- When a safety-related function is defective the residual clinical risk is increased.
- The priority with which defects should be fixed must take account of the clinical risk and the assessment needs to take place at the optimal point in the development lifecycle.
- Where a system goes live with unfixed safety-related defects, these need to be spelt out in the safety case along with any mitigating actions which users or organisations need to take.

References

1. IEEE Computer Society. IEEE Standard 829–1998, "IEEE Standard for Software Test Documentation". 1998.
2. Institute of Medicine. Health IT and patient safety: building safer systems for better care. Washington, DC: National Academies Press; 2012.
3. International Organization for Standardization. ISO/IEC 25010:2011 Systems and software engineering – Systems and software Quality Requirements and Evaluation (SQuaRE) – System and software quality models. 2011.

Chapter 17
Gathering Evidence for the Safety Case

Evidence is an essential part of the safety case, in fact without it its claims can be considered invalid. Evidence comes in many forms as has already been pointed out in this book. The challenge for safety case authors is gathering that material together in a way that justifies its claims without overburdening the reader with detailed data. Using test evidence to support the safety case has already been set out in Sect. 16.3. This chapter turns to other sources and looks at how evidence can be set out in a manner which is compelling to the reader.

17.1 Evidencing Design Features

Much of the bulk of the hazard register will illustrate the system features which have been implemented to mitigate specific hazards or causes. But a simple description of the functionality in the hazard register might not be convincing enough to a critic of the safety case. This is particularly so for those system features which either mitigate high levels of initial risk or are heavily relied upon to reduce risk. In these cases the argument in the safety case should be developed further so that a compelling story is set out leaving the reader in no doubt that the risk has been reduced to as low as reasonably practicable. For lesser risks a combination of requirements, engineered controls, test material and some traceability is sufficient to evidence risk reduction. For more important controls the tools required are narrative, description, logic and story-telling.

The story we set out in the safety case might follow this kind for logic:

1. The requirements were assessed and we found some potential hazards…
2. To mitigate these hazards we designed the system so that it did this…
3. We needed to make sure those things worked properly so this is how we tested them…
4. And when we did that testing we found that…
5. So in summary, this is why we think the risk has been mitigated…

A. Stavert-Dobson, *Health Information Systems: Managing Clinical Risk*, Health Informatics, DOI 10.1007/978-3-319-26612-1_17

There could be many variations on this theme depending on style and subject matter. The point is that the approach is simple and all in one place. As a reader I don't need to wade through test plans, defect logs, requirements matrices or any other collateral – those artefacts are available as and when I want to see further evidence. Where there are multiple key hazards this kind of story would be repeated, not for every last cause and control in the hazard register but for those where the risk justified the effort.

17.2 User Interface Evaluation

Section 5.4 discussed the potential for a system's user interface to contribute to a hazardous environment and therefore to clinical risk. For products reliant on a user interface for their operation it is therefore prudent to undertake an evaluation to determine whether the user interface successfully mitigates human factors.

The extent to which usability can impact safety has long been established and the design and evaluation techniques involved are well documented in the literature [1, 2]. For medical devices, there is usually an explicit regulatory requirement to execute an evaluation which often translates into complying with Standard ISO 62366:2007 [3] or specific FDA requirements [4]. These approaches may be overkill for HIT which falls short of regulation. Nevertheless a great deal can be learnt from these requirements and their methods be applied to an extent commensurate with the risk to non-medical device HIT products.

Usability evaluation can essentially be seen as happening in two phases:

- Formative evaluation – which occurs during the development process and ensures that appropriate appraisal and expertise feeds into the design and that suitable improvements are made, and
- Summative evaluation – which is a validation exercise to determine whether the product that has been built meets usability requirements and the needs of the system's users.

For the safety case, evidencing both evaluations is particularly convincing especially when carried out by the manufacturers. The formative evaluation helps to shape the requirements, take account of control options, iteratively refine the solution and build confidence in the design strategy. The summative evaluation allows specific user interface controls to be validated both in terms of their presence and effectiveness.

A full evaluation of the user interface will look at a number of characteristics including:

- Efficiency
- Effectiveness
- Cognitive load
- Ease of learning
- Ease of remembering
- User appeal

With such a range of metrics it needs to be clear in the test plan whether the intention is to evaluate usability generally or whether to focus on safety. Where safety is the objective the fundamental evaluation process will not change but the weighting one might place on certain types of defects might. For example, whilst shortcomings in efficiency and user appeal might affect the product's marketability, business benefits and staff uptake, these issues are less likely to introduce hazards. For safety the key metric is effectiveness, i.e. the extent to which the user interface supports an individual in accomplishing their objectives. To this ends the evaluator is looking for cases, and in particular consistent cases, of performance failures. These should include scenarios where the user experiences confusion or misinterpretation but manages to recover before harm occurs.

17.2.1 User Interface Evaluation Methods

Traditionally usability evaluation methods are divided into three categories:

Inquiry – whereby an assessor interrogates potential users to characterise their preferences, establish their design requirements and observe them using the system in their actual working environment.

Inspection – whereby usability or subject matter experts undertake a methodical review of the user interface and design material.

Testing – whereby representative users are observed working through selected system tasks whilst an assessor records how well the user interface actually performs.

These approaches are not mutually exclusive but rather complementary revealing different aspects of usability at various point in the product lifecycle. In practice an evaluation cannot investigate every function and every screen in the system. Assessment therefore needs to be a targeted activity, in the case of CRM focussing on those screens and functions where the user interface has the potential to contribute the greatest risk.

There are many formal methods for user interface evaluation in the literature. A selection of some of the more common approaches which are suitable for HIT are set out below.

Usability Interviews

Whilst interview and questionnaires lack the objectivity of formal usability studies, at the early stages of a project they can be applied with great effectiveness to understand the user's requirements both for the user interface and general design. These techniques have strengths in information gathering and quality design rather than as a specific evaluation tool. A number of formal methods have emerged in this area such as contextual inquiry and contextual design.

People, however good at their job, often find it difficult to articulate precisely how they execute their role, how they make decisions and how the environment around them affects their work. Contextual inquiry is a technique which researchers can use to gain a much better understanding of the challenges potential users face. Rather than interviewing a worker in a sterile, unfamiliar setting taking valuable time out of their working day, the researcher shadows the subject whilst they perform their role in the real world. Essentially the researcher takes the role of apprentice noting what the target individual does, how they interact and what decisions they make. From time to time the researcher may ask questions to seek clarification and validate the observations noted.

The results from multiple interviews are collated and common themes established. This information is harnessed to guide the design of a system and, importantly, its user interface through a process called contextual design [5]. The data from individuals under study is formally modelled and consolidated before being delivered to a design team who develop a vision for the solution. The solution is articulated through storyboards and prototypes and validated with the original interviewees. The objective is to create a product which closely meets the user's requirements both in terms of the needs they articulate and, importantly, those which are tacit.

Expert Review and Heuristic Evaluation

In the expert review process a suitable expert is provided with access to the target system or prototype and their view on the product's usability strengths and weaknesses is solicited. The review can occur at any point in the development process and whilst the technique is not a substitute for formal usability testing it is simple, structured and easy to learn.

The choice of expert or experts is crucial in obtaining a meaningful review. Options will include:

- Usability experts
- User interface designers
- Cognitive psychologists
- Clinicians
- System trainers
- Business analysts
- Writers

The results of course may differ depending on the individuals chosen and a group with a range of expertise might provide the best outcome. One of the characteristics of the technique is that generally the experts are not the intended users of the system. Instead they are individuals with experience and expertise in critically evaluating user interface designs.

Each expert is provided with enough training to allow them to carry out the review. They are asked to individually walk through each target screen noting areas of concern. Issues are logged and attributed a level of severity and at the end of the review a session leader collates the responses. Generally the experts are not allowed the freedom to collaborate prior to submitting their review so that a degree of independence is preserved. Finally the leader summarises the findings in a written report ready for engineers to action.

Rather than ask the assembled experts to subjectively comment on the user interface a more objective and quantitative approach is generally called for. Assessors are typically asked to compare the performance of the user interface with a set of usability heuristics. Heuristics are ‘rules of thumb’ rather than specific requirements. They give some idea of the broad principles of good design without being prescriptive or suggesting specific solutions. Despite being relatively generic in nature, design rules like these have found to be invaluable in providing a useful stimulus for critical evaluation.

The most prolific set of usability heuristics in common use are those described by Nielsen in his 1994 publication Usability Engineering [6]. These have been successfully adapted and applied to amongst other things the evaluation of medical devices [7]. Heuristics have also been developed for use specifically with Electronic Health Records [8].

Expert review and heuristic evaluation has a number of limitations. For example:

- The assessors might be experts in design or usability but fundamentally they are not real users of the system. Actual users are likely to have a different set of skills, come from a range of backgrounds and perhaps have a different level of understanding of the system.
- Evaluators typically undertake the exercise in laboratory conditions rather than a workplace setting devoid of the natural interruptions which are so common in clinical practice.
- Often evaluations take place on prototypes which may offer limited integration.
- The technique is poor at identifying missing functionality.
- The method fails to provide recognition of those components where the design correctly follows the guidelines.

Cognitive Walkthrough

Cognitive walkthrough (CW) is an inspection technique developed by Wharton et al. [9] which focuses on learning through exploration of the system. It is a particularly useful method for systems where the user has little or no previous experience or training (such as a kiosk or consumer-facing website). The method has however been applied to a wide range of systems in many different industries.

CW is performed by expert evaluators either individually or, more commonly, as a small group. In preparation for the evaluation, the development team constructs a series of realistic tasks or ‘user goals’ with a clear start and end point (for example, initiate and commence a telemedicine consultation or find and cancel an existing appointment). When safety is the focus of the exercise, the team should choose tasks which, on the basis of the hazard register, are known to be associated with significant risk. Each task is broken down into a set of action sequences which lead the evaluator through the system workflow to the end point.

The evaluators are presented with the task under study and asked to work through each individual action. The evaluator examines the user interface and executes the next action they believe will progress towards the defined goal. At each step individuals are asked to note any difficulties or areas of concern. The evaluator records either a success or failure story depending on whether they were able to successfully progress.

At the end of the session the results are collated and the evaluators asked to summarise the experience. The audience may be asked to suggest potential improvements to the user interface or solutions for the problems they encountered. The facilitator makes careful note of the success and failure stories reported by the evaluators for further analysis with a view to improving the interface.

The technique is simple to learn, and cost effective. Although the method requires access to the user interface this can be a prototype so the evaluation can still take place reasonably early in product development. As with heuristic evaluation the technique suffers from being executed by individuals who are not typical users – indeed it can be difficult for designers to deliberately act like users. To combat this the method can be augmented by developing a persona or user profile outlining the typical kind of user, their intentions, working environment, motivations, distractions, etc. This rich narrative is then presented to the assembled group prior to the exercise and taken into consideration as part of the evaluation.

Usability Testing

The techniques listed so far have been largely formative in that their strength is in teasing out inadequacies of the user interface in order to drive improvement during the design and build phases of a project. But for the safety case, it is the summative evaluation which largely provides the specific evidence needed to justify its claims. Tool such as User Acceptance Testing are invaluable in validating that a system has the right functions but says little about how it actually performs when operated by real users. For good reasons functional testing is tightly scripted but to truly validate that the solution is fit for purpose requires the relative freedom and judgement calls that usability testing permits. Importantly, the purpose should be a genuine and honest exercise to gauge objective opinion rather than a railroading of participants into signing off the product and/or elements of the safety case.

17.2.2 Undertaking Usability Testing

However well informed a design process may be, ultimately the developers need in the course of designing the user interface to make assumptions about the users' workflow, their environment and objectives. Those assumptions may turn out to be valid or invalid and it is the process of usability testing which puts that to the test. Unlike inspection and inquiry, formal usability testing examines the product itself or at least a mature prototype. It is therefore usually undertaken later in the product lifecycle but of course not so late that modifications are impractical or costly.

Usability testing should be a planned activity with defined pass and fail criteria and the strategy should be set out in a test plan or similar artefact to be agreed up-front by project stakeholders.

Usability testing begins with the recruitment of suitable participants. The subjects should represent the typical intended user groups; clinicians, administrative staff, system administrators, trainers, technical staff, etc. – personnel involved in developing the product are generally excluded. Ideally subjects should be chosen from a suitable range of care settings and it should be made clear to participants that it is not they who are under scrutiny but the system. The testers are likely to need a degree of training or already be proficient at operating the system. The level of training needed is a judgement call but remember that training beyond what would be provided to ordinary users might lead them unduly and compromise the detection of learnability issues. Some strategies deliberately include 'worst-case' users who are untrained or have limited language proficiency. Each testing session should be led by an individual experienced in usability and human factors evaluation.

Usability testing can be undertaken in either laboratory conditions or in the real world. Either way the intention should be for testing to be done on an environment which is as close in look and feel to the live platform as possible. The system should be configured in a realistic manner and be populated with appropriate test data which is representative of the clinical setting.

Evaluators need to decide at the outset which system workflows will be examined based on the potential clinical risk. The number tackled at each session should be realistic to allow time for note taking and feedback. If the primary purpose is to validate the hazard register then the tasks under examination should clearly map to those hazards reliant on human factor controls.

The session begins with a briefing on the objectives and methods and a summary of the tasks scheduled for examination. Each participant then works through the assigned tasks to reach a designated end point. They are asked to note any difficulties, inconsistencies or peculiarities they might observe. The following should be carefully recorded in the course of testing:

- The number of errors made at each stage of the process
- The severity of the errors and whether in the real world those deviations could result in a hazardous scenario
- The different paths and dead ends the user investigates before accomplishing their goal
- The number of times the user has to consult help material or ask for assistance

To bring objectivity to what can at times seem like a subject evaluation, testers can be provided with a review checklist based on the product's user interface requirements or a set of suitable heuristics such as Neilsen's [6]. The evaluator should also be diligent in recording subjective feedback and it can be useful at the end of the session to make a point of gauging general opinion and consensus. Some studies make use of video recording or input logging systems to detect subtle errors in interaction, hesitation and recovered errors.

A popular variation of usability testing is a concept called 'think-aloud' introduced by Clayton Lewis at IBM. With this model a participant is asked to vocalise what they are thinking, doing and feeling as they work through the designated tasks. An evaluator notes what is said or the session is recorded for later analysis. The protocol is sensitive to detecting when the user is surprised, confused or frustrated by the system, the continuous commentary providing valuable context and insight into what is going on in the subject's mind. Unfortunately not all individuals are equally adept at vocalising their thoughts whilst operating the system and well-trained evaluators are needed to offer timely prompts and reminders in the course of the test. There is also the potential for subjects to be unduly influenced by the evaluator or other circumstances of the test.

Once the results of usability testing are collated, the evaluator should author a brief report outlining the findings. The observed defects should be attributed a level of priority and it can be useful to develop word pictures to describe each category (e.g. Serious = User is unable to continue the workflow). A template for the report can be helpful, for example, The National Institute of Standards and Technology have developed a design review protocol specifically for evaluating and documenting Electronic Health Record reviews [10].

Each defect needs to be subject to a CRM assessment and a level of clinical risk assigned to each to inform the fix priority. It should be possible to map each defect back to an appropriate hazard in the hazard register. If this is not possible then the team should discuss whether in fact a previously unforeseen hazard has been detected.

Evaluators should particularly be on the lookout for defects which might have only been spotted once but could affect large parts of the system. For example, suppose a tester notes that the 'Okay' and 'Cancel' buttons are too small to be used on a mobile platform, this single comment could represent an endemic problem across the entire solution and should be prioritised accordingly.

Finally, effectiveness is not the full story when it comes to usability and safety. Section 5.4.2 discussed the notion of cognitive load and the potential for unwieldy user interfaces to place excessive demands on users' brainpower and contribute to slips and lapses. To this extent a formal assessment of cognitive load during usability testing can contribute significantly to evidencing that this has been considered in evaluating user interface hazards. There are many methods for assessing cognitive load but often they require access to a cognitive psychologist or expert in the technique. However there are a small number of established approaches which make use of simple questionnaires, for example the NASA Task Load Index (NASA-TLX). There has even been some success in using this to evaluate user interfaces in healthcare [11].

17.3 Validating the Training Programme

We discussed the use of system training as a control in Sect. 15.3. Whilst human factors generally represent controls at the least effective end of the spectrum many systems rely on the behaviour of users to mitigate risk to some extent. Training is one way by which users can be encouraged to adopt consistent behaviours and develop a culture of best practice. If one intends to justify in the safety case that risk has been mitigated to acceptable levels then it is useful to set out the rigor which has been applied in developing in the user training programme. In fact it is the very fallibility of human factors as controls which requires us to make optimal use of the few tools we have available to make them more effective.

17.3.1 Validating Specific Training Controls

The hazard register is likely to contain at least some controls which are partly or entirely dependent on users operating the system in specific ways. Providing evidence that these dependencies have been addressed during training can support the safety case's claims significantly. Of course readers will only have confidence in the report if the evidence is compelling. For that one needs to demonstrate that (a) key safety messages were actually delivered to users and (b) that this resulted in users exhibiting the correct behaviour during system operation.

In Sect. 15.3.1 we discussed how controls in the hazard register can act as the seed for highlighting key concepts in the training programme. This linkage between human factor controls and training material is important beyond the practical creation of the course content. It provides safety case authors with a means to evidence the real-world implementation of the controls and, importantly, to highlight any gaps. If gaps are found then appropriate corrective actions can be taken and this too evidenced in the safety case.

For example, suppose one has a control as described earlier:

> Users should be made aware that when a patient has more than five recorded allergies, it will be necessary to click on the ellipsis '…' to reveal the remaining details.

Implementation of this control might include evidence as follows:

> - Training Programme Exercise 3.7 requires learners to discuss the potential hazards of relying on the system to present a complete list of allergies. This issue is highlighted in the facilitator's notes to ensure it is raised during the discussion.
> - Step 3 of Training Scenario 7 (Reference TSS007) highlights the need for users to click on the ellipsis and the learner is required to do this in order to complete the scenario.
> - Standard Operating Procedure 4.5 (Review of a patient's allergies prior to administering treatment) emphasises the need to take this action in Step 6.
> - Test data has been set up in the training environment to ensure that users will actively be presented with patients who have more than five recorded allergies.

Whilst evidence of this kind supports the claim that a control has been put in place it does not necessarily demonstrate that the control is effective. To do this one must demonstrate that users actively comply with the policy and that as a result of complying, the risk of a hazard being triggered is reduced. Depending on the control in question, this is often challenging to evidence in HIT systems especially before operational experience has been gained.

In some cases it may be possible to create a test or database query to detect whether certain procedures are adhered to in live operation. For example, in this case one might be able to technically check how often the query to retrieve all allergies is executed when a patient has more than five entries recorded. Often the most informative analysis occurs not from the value which the query returns but as changes and trends are observed over time. Should today we observe that the all allergies are viewed 70 % of the time but last month it was 90 % this might suggest that lapse errors or violations are increasing in prevalence and that a degree of re-training is required. The technical means to achieve this monitoring and a documented plan to periodically execute it constitute evidence for the safety case. As always the time, effort and cost of undertaking this level of vigilance should be balanced against the risk of the hazard being mitigated.

In some cases it will not be practicable or even possible to actively monitor adherence to operating policy. Instead one may need to undertake a degree of retrospective analysis by looking at reported adverse clinical incidents. For example, one might choose to investigate those reported incidents which relate to a patient being administered a medication they were allergic to. Should it be observed that a disproportionate number of patients had more than five known allergies then this may be suggestive of an underlying issue. Again, trends and patterns are often more helpful than absolute numbers. A retrospective approach of this kind can be labour intensive and of course by its very nature fails to proactively reduce risk. Nevertheless important lessons can be learned and effective incident reporting and analysis can act as a backstop for evaluating the effectiveness of some important controls.

17.3.2 The Training Approach as Evidence

For systems which rely significantly on human factors and training as controls it can be useful to set out the general approach to training in the safety case. This form of evidence alone is generally insufficient to demonstrate that specific controls have been mitigated, however it does provide a reader with an appreciation of the diligence which has gone into providing a quality learning experience.

At the most fundamental level, training as a mitigator of risk can only be effective if users of the system actually participate in training. If only a proportion of users are trained at the point of go-live, training related controls cannot be relied upon and this is likely to have a significant impact on the residual risk. Indeed, thought should be given as to whether users should physically be allowed access to the system if they have not completed the relevant training for their role and access permissions. Finding a convenient time and place to train clinical staff can be a particular challenge. Clinicians have to deal with many conflicting demands on time which often change at very short notice. For these reasons it is particularly important that the extent of training coverage can be evidenced in the safety case. Healthcare organisations need to keep detailed records of who completed training and when. Each iteration of the safety case should summarise the current picture along with a plan to meet any training gaps. If training coverage is low then a risk assessment should be undertaken and the impact of incompletely mitigating certain hazards analysed.

Other characteristics of the user training programme which can help to demonstrate training diligence and bolster confidence in the safety case include:

- The ongoing provision of easily accessible documentation to support training including Standard Operating Procedures, user guides, video tutorials and application help functionality.
- Traceability between course content and the functional scope of the product to evidence that all relevant system functions are adequately covered by the course.
- A training needs analysis to demonstrate that individuals are provided with the necessary training to meet their specific role.
- Involvement of the system supplier to ensure that a baseline level of application knowledge exists amongst local trainers and to act as a reference point when support for training is required.
- The provision of a system training environment and appropriate hardware configured in the same way as the live system and populated with realistic test data for learners to practise operating the system.
- The presence of a process to identify new members of staff and provide them with training before they operate the system.
- A strategy to repeat training on an ongoing basis in order to re-enforce good practice particularly when safety-related changes to the system are envisaged.

17.4 Validating Operating Policies

Section 15.1 discussed the relative effectiveness of controls and the idea that those which depend on human factors tend to be less reliable. Nevertheless effectiveness can be improved by employing active measures to validate the presence and effectiveness of these controls. For some human factor controls it may be appropriate to evidence their implementation through training provision alone as set out in the previous section. But for hazards at the higher end of the risk spectrum this strategy may be insufficient.

Increasingly in healthcare protocols are being implemented to guide clinical decision making and business workflow. Protocols bring consistency to healthcare delivery and the sharing of best-practice. Whilst some are formulated as guidelines to be implemented at the clinician's discretion (sometime referred to as passive protocols) others (active protocols) contain formal prescriptive rules. The UK's Department of Health publication Organisation With A Memory recommends that "consideration should be given to the production and piloting of standardised procedural manuals and safety bulletins which it is obligatory to use when embarking on specific high risk procedures" [12].

The notion of working to agreed rules to mitigate risk is therefore not unfamiliar to healthcare professionals. Where rigour in control implementation is called for, formal artefacts such as operating policies need to underpin how we expect users to behave – essentially defining the rules of engagement. Policies of this nature are usually formulated by healthcare organisations and take into account local procedures, configuration and workflow. Operating policy development ideally needs to be completed before the system is put into live service and then be subject to regular review and update.

For any operating policy to be successful it must:

- Contain a step by step account of what the user needs to do
- Be clearly written covering all reasonable scenarios in which the user may find themselves.
- Explain to users why they need to behave in particular ways to reduce risk rather than blindly dictate the rules
- Be agreed by a representative sample of those users who will operate within it.
- Be up to date and accessible by those who need to use it particularly when staff join the organisation or team.
- Be reasonable so that what it asks of users is workable and not over-restrictive or grossly inconvenient.

In terms of evidence, the safety case should set out (for those key hazards where human factors are important mitigations) a clear chain of logic from the control to a workable and effective policy. At the most basic level this should consist of a formal reference to the document or traceability from a series of controls to several relevant policies. By supplementing this with a description of how the above policy characteristics were achieved, further confidence can be built to show that appropriate diligence and governance has been applied.

To provide a further level of assurance one can undertake a formal review of those operating policies which are critical for maintaining safe operation. The outcome of the review can then be documented as evidence in the safety case. A checklist of desirable policy characteristics would be one way of demonstrating that a methodical approach had been employed in the review.

In the real world individuals don't always abide by the rules. It has been shown that healthcare professionals have a tendency to forget to follow protocols or deviate from them even when they are available [13]. With this in mind it is prudent to take reasonable measures to ensure that key operating policies are complied with. Hudson et al. [14] provides a useful list of steps to reduce policy violation. These can be translated into a helpful set of prompts to evidence a policy review:

1. Are the rules correct, comprehensible and available to employees and understood by them?
2. Have any unnecessary, out-of-date or unworkable procedures been revised or abandoned?
3. Is compliance with the policy actually possible given the training, equipment and time available?
4. Have any factors which encourage the violation of rules been eliminated?

17.5 Building Confidence in the Safety Case

Some risk management activities stop short of validating specific controls. They do however provide the reader with confidence that potential faults will have been found, that care and attention has been applied in the design and build and that a level of quality assurance has been provided. These matters deserve their place in the safety case and add weight to its claims.

17.5.1 Historical Operation

Section 14.1.6 discussed how novelty in design and implementation can be a common source of potential risk. Conversely, previous successful operation of a system has the potential to provide a source of risk management evidence for the safety case. In those first implementations teething problems will have been identified, potential hazards are likely to have been characterised and the effectiveness (or otherwise) of controls established. The manufacturer is likely to have learned lessons from users of the system and vice versa. For healthcare organisations occasionally there can be practical challenges in obtaining the sufficiently detailed information from the manufacturer to be used as evidence. In fact the contractually enforced non-sharing of known issues across organisational boundaries is unfortunately widespread [15].

Nevertheless, rich information from existing customers can sometimes be obtained such as:

- Already developed operating policies and training material
- Established hazard registers
- On-going issue logs and work-off plans
- Current defect list
- The views and general experience of users from existing sites

Note that it does not necessarily follow that previous use of a system equates to safe operation and risk reduction. Indeed after go-live the risk of a product may be assessed as being greater once it is tried and tested in the real world. Equally, thought should be given as to how similar previous implementations are to that under consideration. Should there be significant differences in intended purpose, care setting, user numbers, technology or architecture any like for like comparisons in the safety case could be subject to criticism.

17.5.2 Conformance with Standards

Conforming to specific standards does not directly translate into a safe product. Nevertheless where standards exist in the subject matter being studied to turn a blind eye to them without justification could be folly.

Standards can help to manage risk in HIT in a least a couple of ways. Firstly they share best practice. Risk can never be eliminated but following best practice at the very least demonstrates an organisation's commitment to doing the right thing. Should the system's safety profile ever be subject to scrutiny by an external body, showing that state of the art techniques and skills were employed in its design and implementation can go a long way to diffuse criticism. This is particularly true for those standards which relate directly to safety and quality management (see Chapter 4). Having said that complying with one or more safety/quality standards rarely provides evidence for the mitigation of specific hazards.

Secondly standards can help to forge consistency. Occasionally hazards come about not because a workflow or user interface is suboptimal but simply because it differs from what individuals have been used to (see Sect. 5.4.3). By following standards one would hope that the design would benefit not only from best practice but also from consistent practice.

In some cases such as inter-system messaging, compliance with agreed standards such as Health Language 7 (HL7) is a functional necessity. Note though that simply supporting a standard such as this is not necessarily evidence for the safety case in itself. What matters is the extent to which conformance with the standard has been assured. If one were to verify through static and dynamic testing that the actual messages formed were as predicted and that integration testing was successful in gaining communication with an external system already HL7 compliant then this is far stronger evidence than citing HL7 support alone.

17.5.3 The Clinical Risk Management Workshop

Sections 13.2 and 13.4 discussed the SWIFT method for hazard derivation and the need for brainstorming. These techniques are often carried out in a workshop environment involving key stakeholders and domain experts. The act of employing this methodical evaluation of the system is important evidence which raises confidence in the safety case and provides non-specific but nevertheless important risk mitigation and diligence. This is particularly the case where structured documentation is made available to support the workshop and a clear set of inputs and outputs are defined.

The safety case can include a brief summary of one or more workshops setting out the key items for discussion and the conclusions which were drawn. A verbatim account simply isn't warranted but providing the reader with the satisfaction that due diligence was followed, the right people were engaged and that all necessary factors were taken into account goes a long way to raising confidence in the report's claims.

17.6 Summary

- Evidence for the safety case can come from many different sources and it extends well beyond software testing.
- Design features intended to mitigate risk often make up strong controls and their verified presence and effectiveness constitute valuable evidence.
- The user interface can be a source of many hazards and poor usability is an issue in many software domains. Thankfully user interface assessment and validation techniques are ubiquitous and well described in the literature.
- Training is generally a less effective control but that doesn't mean its delivery or otherwise can be ignored. Human factor controls can be difficult to evidence so when time and effort are put into a robust training programme its features form useful material for the safety case.
- Operating policies are valuable assets to an organisation and, along with training, provide a means of systematically communicating human factor controls down to user level.
- Readers need to have confidence in the safety case's claims. Whilst evidence of historical operation, conformance with standards and well-managed CRM workshops are non-specific controls they can add up to a rounded and convincing story.

References

1. Lowry S, Quinn M, Ramaiah M, Schumacher R, Patterson E, North R, et al. Technical evaluation, testing, and validation of the usability of electronic health records. NISTIR 7804. 2012.
2. National Patient Safety Agency. Design for patient safety. User testing in the development of medical devices. National Reporting and Learning Service. 2010. Report No.: Ref. 1184.
3. International Organization for Standardization. IEC 62366:2007. Medical devices – application of usability engineering to medical devices. 2007.
4. U.S. Food and Drug Administration. Human factors and medical devices. [Online]; 2015 [cited Aug 2015]. Available from: http://www.fda.gov/MedicalDevices/DeviceRegulationandGuidance/HumanFactors/.
5. Beyer H, Holtzblatt K. Contextual design. Defining customer-centered systems. 1st ed. San Francisco: Morgan Kaufmann; 1997. ISBN 9780080503042.
6. Nielsen J. Usability engineering. San Diego: Academic; 1994. ISBN 0125184069.
7. Zhang J, Johnson T, Patel V, Paige D, Kubose T. Using usability heuristics to evaluate patient safety of medical devices. J Biomed Inform. 2003;36(1–2):23–30.

8. Armijo D, McDonnell C, Werner K. Electronic health record usability. Evaluation and use case framework. Rockville: Agency for Healthcare Research and Quality. U.S. Department of Health and Human Services; 2009. AHRQ Publication No. 09(10)-0091-1-EF.
9. Wharton C, Rieman J, Lewis C, Polson P. The cognitive walkthrough method: a practitioner's guide. 1994.
10. Svetlana L. Customized common industry format template for electronic health record usability testing: national institute of standards and technology. U.S. Department of Commerce; Gaithersburg, Maryland; 2010.
11. Longo L, Kane B. A novel methodology for evaluating user interfaces in health care. 2011.
12. UK Department Of Health. An organisation with a memory. Report of an expert group on learning from adverse events in the NHS chaired by the Chief Medical Officer. 2000.
13. Claridge T, Parker D, Cook G. Investigating the attitudes of health care professionals towards SCPs, a discussion study. J Integr Care Pathways. 2005;9:1–10.
14. Hudson P, Verschuur W, Lawton R, Parker D, Reason J. Bending the rules II: why do people break rules or fail to follow procedures, and what can you do about it? 1998.
15. Koppel R, Kreda D. Health care information technology vendors' "hold harmless" clause: implications for patients and clinicians. JAMA. 2009;301(12):1276–8.

Chapter 18
Developing the Clinical Safety Case

Writing a coherent and concise safety case is an art and requires skills that only come through experience. Structure is vital as without this it is impossible to fluently articulate the safety argument in a logical manner. It is a good idea to draft out the structure first formulating section headings which flow and tell a story. Ultimately the report should draw some clear conclusions on risk acceptability and make recommendations on next steps.

The author of the safety case should be an individual who is capable of drafting articulate text, organising their thoughts and summarising evidence. Note that whilst a clinical individual might ultimately sign off the report it might not be necessary for the author to have either deep clinical or technical knowledge. One could argue that in building the safety case the ability to collate and rationalise information is more important than being skilled in the precise subject matter.

18.1 Safety Case Structure

A possible structure for the safety case is set out below:

1. Introduction and executive summary
2. The safety claims
3. System description and scope
4. Assurance activities
5. Clinical risk analysis and evaluation
6. Clinical risk control and validation
7. Outstanding issues
8. Conclusion and recommendations

A. Stavert-Dobson, *Health Information Systems: Managing Clinical Risk*,
Health Informatics, DOI 10.1007/978-3-319-26612-1_18

18.1.1 Introduction and Executive Summary

This section will normally set the scene and rationale for undertaking the CRM assessment. This serves to put the report into the context of the overall project and product lifecycle with reference to key milestones. It is often appropriate at this point to refer back to the original CRM Plan for further context setting.

The executive summary serves to brief a reader on the salient points in the report without requiring detailed scrutiny of the rest of the document. The summary will typically contain a view on the overall clinical risk profile and acceptability along with any outstanding issues or constraints on the assessment. Key claims and recommendations may also be iterated in this section to round off the summary.

18.1.2 Safety Claims

In many ways the safety argument can be visualised as a hierarchy with detailed evidence at the bottom and the key claims at the top. Claims represent the ultimate goal of the safety project, simple statements that set out what it is that one is striving to achieve. For example, we may choose to claim that:

- All reasonable measures have been undertaken to identify potential hazards, and
- That all identified hazards have been mitigated to as low as reasonably practicable.

Alternatively we may decide to create more specific claims, perhaps that the system performs at a level that is conducive to practical operation in a clinical environment and/or with a defined degree of availability. Whatever measure is selected it should be clear and concise and, most importantly, be capable of being demonstrated within the scope of the planned CRM assessment process. In other words it should be possible to draw lines of logical inference between each claim and the arguments set out in the body of the safety case.

18.1.3 System Description and Scope

In Sect. 11.1 we discussed the importance of defining the system by ascertaining its intended purpose and boundaries. It is worth reiterating this in the safety case itself as by the time this is being constructed it is likely that one will have a better handle on the precise nature of the requirements, functionality and design. Similarly it can be helpful to restate the rationale for the assessment and its regulatory position. As a minimum the text should set out:

- The formal name and version of the product under assessment
- The high level clinical business processes the system will support
- Who the users are and how they will use the system

- The dependencies on other technologies, interfaces and system architecture
- The need for any data migration or transformation activities

In the event that the scope of the system has changed during the course of the project this should be discussed. This is of particular significance where a change has occurred since a previous iteration of the report and a re-evaluation of the clinical risk has been necessary. In some cases the scope may have changed as a direct result of the findings of the safety assessment particularly where an approach of 'control by elimination' has been taken (see Sect. 15.1.1). In describing the system it can be useful to break the scope down into logical areas of functionality and document very briefly the rationale for the component being related to safety (or otherwise). This gives the reader a flavour for the analysis to come and an approximation of the magnitude of risk involved.

Note that in defining the system one should be careful not to use overly 'commercial' language. The safety case is not the place to extol the virtues of efficiency gains, improved productivity, return on investment, etc. These are all important facets of product management, the business case and marketing but in the safety case these factors could be perceived as being an ill-informed means of justifying the clinical risk. Any text which is reused from other materials should have subjective elements removed so that a plain and factual system description is articulated.

18.1.4 Assurance Activities

The safety case should contain some reference to the method that has been used to derive it. Demonstrating a rigorous and structured approach will improve a reader's confidence in the report's claims. A judgement needs to be made on the level of detail which is considered appropriate in the context of the project. In many cases a reasonable approach might be to simply reference other CRM process documentation and the CRM Plan to avoid duplication of the same material. However careful thought should be given as to whether the process documentation will be available to all readers of the safety case. For example, suppose a safety case simply states that the manufacturer's standard CRM methodology was applied to the product and references made to internal process documentation. A healthcare organisation might be within their rights to question the nature and rigour of a process which is less than transparent. This is where setting out a summary of the approach in the safety case can restore confidence without necessarily divulging what might be valuable intellectual property.

In some cases, especially where an organisation is managing a small and single product, it may be more efficient to set out the entire CRM process within the safety case itself rather than maintain this information separately. This has the advantage that the safety case becomes a self-contained justification without requiring the reader to consult extraneous material.

Section 11.1.6 introduced the notion of assumptions and constraints on the assessment. It can be valuable to revisit these in the safety case as they may well have changed since the original CRM Plan was authored. In many cases assumptions will have been proven valid or otherwise and constraints may have been overcome. The safety case should summarise the current position and set out how those concerns were managed during the lifecycle of the project along with any impact on the clinical risk.

18.1.5 Clinical Risk Analysis and Evaluation

The risk analysis and evaluation will form the heart of the safety case. It is here where the argument and evidence is set out to justify the report's safety claims. This information is inextricably linked to the detail in the hazard register but in the safety case one has the opportunity for further explanation and elaboration. The hazard register typically has a formal structure to support cohesion and consistency between hazards. The free-form text of the safety case facilitates articulation and openness enabling us to support our argument with direct and indirect evidence. Essentially the text provides the necessary inference between the hazards, causes, controls and evidence that is needed to justify the argument.

Careful consideration needs to be given to the breadth and depth of detail one decides to include in the documented analysis and evaluation. At one extreme there is a danger of reiterating the entire hazard register in this section of the report. There is usually little to be gained from this approach as the hazard register is usually included along with the safety case anyway. Pages of text justifying the mitigation of a hazard from low clinical risk to very low almost certainly fails to warrant the effort involved. In an overly elaborate report it can quickly become impossible to tease out those hazards, causes and controls of most significance. In contrast, insufficient analysis and evaluation can result in a bare report which fails to lead the reader through a logical path of justification.

A useful exercise is to develop criteria for significant hazards or areas of functionality worthy of specific examination. Subject matter meeting the criteria is promoted for explicit discussion in the safety case. For example, one may consider developing the argument in detail where:

- The initial and/or residual risk is above a certain level
- The evidence to support the safety argument is complex, indirect or lacks transparency
- It is necessary for external stakeholders to implement specific controls
- The measures employed to mitigate the risk to acceptable levels are extensive or complex
- There is a particular reliance on human factor controls
- There are outstanding risk reduction activities which have yet to be completed

The following could be a fragment from an order entry system safety case. It briefly sets out the problem, a series of controls and some evidence that the controls were implemented correctly. The text is by no means complete and more detail would provide significantly more confidence but as an early iteration it presents some basic argument and evidence.

During initial evaluation of the user interface design it was identified that there was the potential for a user to select an incorrect investigation from the long list of tests presented (Hazard 27). This could contribute to a user selecting the wrong item and order an incorrect investigation with a potential delay in clinical care. The issue was evaluated as presenting a Medium Clinical Risk.

Users worked with the manufacturer to enhance the order selection screen to include the ability to search the list by test name and also to select from a list of 'favourites'. In addition users reported that the long list of orderable tests included many which were clinically superfluous. As such, the order catalogue was modified such that unwanted investigations no longer appeared to the user for selection. Finally a request was raised to update the user guide and training material with a warning for users to check that the selected investigation on the confirmation screen is correct before submitting the order.

The system was tested and no defects were reported in the new 'search' or 'favourite' functionality. The project team validated the system order catalogue against the laboratory catalogue and confirmed that superfluous investigations had been removed. The user guide was reviewed by the hospital's training staff and an appropriate warning has been included on page 24 in bold type. The residual risk was re-evaluated and found to be Low.

18.1.6 Clinical Risk Control and Validation

Having set out the argument and evidence in the previous section one might query why it is necessary to include an additional discussion on controls specifically. In some cases further elaboration may simply not be required but in most cases it is useful to have an opportunity to highlight groups of controls and validation activities which support the safety case. In this way one can set out those high level controls which fundamentally manage risk across the entire system. This approach provides a general level of assurance and confidence in the report. Cluttering the hazard register with multiple references to generalised risk reduction activities is likely to create unnecessary noise which can be avoided by including a simple summary in this section of the safety case.

For example:

- The undertaking of system testing, defect identification and fixing is an activity which is likely to reduce risk across the board. This section of the report is also a good location to summarise the outcome of system testing and corresponding risk analysis (as discussed in Chapter 16).
- In the hazard register we may have established a number of disparate hazards all of which require users to be trained to operate the system in a particular way. It can be useful to set out not only the evidence for mitigating each control but also the general methodology for ensuring that the relevant controls are incorporated into the training materials. This might be augmented with a discussion on training attendance records, approach to system administrator training, etc.

- At the design stage a manufacturer may have identified the need to collaborate with users more closely. Perhaps they choose to convene a group of expert users to guide screen design, workflows and operational rules. The convening of these important consultations are likely to reduce risk and a summary of the sessions should be set out in the safety case preferably with reference to formal minutes.

No safety case can be complete without the inclusion of the hazard register. Remember that as the hazard register is continually evolving during the life of the product its inclusion in the report can only represent a snapshot at a point in time. It can be useful to point this out in the report and to ensure that the date and/or version number of the hazard register is indicated.

Note that where hazard registers are large it might be easier in the safety case to refer out to a dedicated document, appendix, annex or electronic report. Whatever format is chosen, it should be complete, intuitively structured and be readily accessible to the intended stakeholders.

18.1.7 Outstanding Issues

Establishing completeness of analysis is a key objective of the safety assessment. Where there are activities to complete or information still to be ascertained these constraints on the analysis should be explicitly declared in the report along with expected resolution, timescales and potential impact on clinical risk. In some cases it can be useful to set out what the pessimistic world might look like should those outstanding activities fail to complete.

18.1.8 Conclusion and Recommendations

The safety case should be brought to a close with a clear set of conclusions which can be logically inferred from the analysis. In particular:

- Does the evidence support the claims of the safety case?
- What is the overall residual clinical risk presented by the system?
- Is that degree of risk is acceptable when tested against the agreed criteria?

Finally the report should specify one or more recommendations for the project. This might be as simple as proposing that the HIT system can progress to go-live or is fit for further roll-out. Recommendations might include the need for future CRM activities, expectations of external stakeholders, further iterations of the document or even caveats to the advice provided. Above all else, any recommendations should be specific, crystal clear and utterly unambiguous to all stakeholders.

18.2 The No-Go Safety Case

In the course of undertaking an assessment it is common to find a need for additional controls and for these to be implemented in the course of the project. From time to time a project milestone may be reached and it be concluded at that point that the clinical risk is either unacceptable or that the analysis could not be completed to a point where a conclusion could be drawn. Occasionally this can come as a surprise to stakeholders and necessitate re-planning or re-engineering late in the development lifecycle. A report of this nature is likely to severely impact the project perhaps bringing it to a halt, significantly extending timelines and/or introducing additional costs. The no-go situation needs careful management and close cooperation of stakeholders and senior management.

Where the outcome of an analysis is unexpected at least some consideration should be given to the possibility that the safety assessment process or its application could be flawed in some way. As part of the action plan it can be prudent to check that the issue is not a figment of the assessment methodology, especially when a CRM process is new and has not been previously tested. For example:

- Have all existing controls truly been established and taken into account in deriving the clinical risk?
- Are the causes of significant hazards actually credible?
- Are the definitions of severity and likelihood categories appropriate?
- Is the appetite for risk acceptability overly pessimistic?

Once these factors have been ruled out one is left with the scenario where the clinical risk associated with the HIT product is genuinely unacceptable and an appropriate action plan needs to be drawn up. The first decision may be to determine whether or not the safety case should be issued at all – indeed, is the safety case a case at all where it concludes that the risk is unacceptable? In general a report of this nature can be particularly enlightening. An opportunity to formally set out the specific areas of concern provides focus and mutual understanding of the problem. A report recommending further work should therefore not be buried but instead form a sound basis for option development.

In some cases one further option might be considered when faced with unacceptable risk. Some standards such as ISO 14971:2012 [1] and the UK standards ISB0129/0160 [2] allow for a further level of justification in the event that the risk is judged to be unacceptable and further risk control is impractical. In these circumstances one is able, through a process of literature review, data gathering and expert consultation, to weigh the clinical benefits of the system against the assessed residual risk. In this way it might be possible to argue that, despite the risk, the overwhelming clinical benefits of the technology justifies putting the product into use. Organisations taking this approach would be wise to take external risk, regulatory and legal advice before proceeding.

18.3 Safety Case Staging

The safety case can be used for different purposes at different times of the development lifecycle. For example, at design time stakeholders can use the identified causes of hazards as a basis for creating effective design controls. During the course of the project the safety argument is developed as the design becomes more real and tangible. In live service, the hazard register and safety case remains an important on-going reminder of the need to support and monitor human factor controls. Failure to provide early sight of the safety case can mean that opportunities are lost to influence the design, testing, operating policies and training material. As such, it can be useful to issue versions of the safety case at key decision points and project milestones. Within an agile development framework, iterations of the safety case may be created on a very regular basis. This iterative approach ties together and evidences the safety activities of the stakeholders in a single referential place.

The intention to develop staged reports should be set out in the CRM Plan for the project. For example, one might choose the following schedule:

- Version 1 – issued during the design phase
- Version 2 – issued after system testing is completed and prior to go-live
- Version 3 – issued 6 weeks after go-live
- Version 4 – issued prior to decommissioning

Note that it can be particularly helpful to reissue the safety case shortly after go-live (Version 3 in this example). This provides the safety case developer with the opportunity to monitor immediate post-go-live issues, often a time when many functional and non-functional defects are recognised. The identification and resolution of these issues can be concisely wrapped up in a post-go-live report with any newly established hazards, causes or controls neatly highlighted. This version of the report will often carry the system through live service perhaps requiring only minor updates as and when changes are implemented or when new information comes to light.

Of course careful thought needs to go into deciding who the staged reports are actually issued to (especially in the absence of any regulatory body). For healthcare organisations the safety case is likely to be made available as an internal document for reference purposes and might be shared with healthcare governing bodies. Manufacturers should aim to issue staged safety cases to their customers and might be included as part of the system acceptance criteria.

18.4 Language and Writing Style

Those who have practised in the world of CRM and HIT will quickly recognise the impact of the 'safety-label'. A rumbling system issue can fly under the radar for quite some time but when the words 'safety' or 'clinical risk' appear, the issue can quickly become the focus of attention and sometimes an outpouring of unhealthy emotion. Of course in reality attaching a safety label means little or nothing. Most issues in the clinical environment have the potential to impact care to some extent and risk is not black or white. It is the degree of clinical risk when formally assessed according to an agreed process which matters and not any kind of binary safety label.

Developing an objective approach to risk management is as much about the language we use as it is the processes we put in place. Those who have the authority to undertake a safety assessment have a responsibility to wield that power carefully and shrewdly and this can quickly be undermined when one resorts to emotive and reactive language. This is as true for the language of the safety case as it is in the corridor conversations with colleagues. Those who operate in CRM have a duty to propagate objectivity by example and to communicate in a way that drives a safety culture which is not rash but considered.

The power of the safety argument comes from one of objectivity rather than subjectivity. So what do these two terms mean?

18.4.1 Objectivity in Clinical Risk Management

Objectivity is the ability to present something on the basis of fact. Subjectivity presents a situation based on opinion, feelings, emotion and beliefs. For example, one could look at a painting and describe it objectively or subjectively. An objective view might describe the painting as portraying a middle-aged woman wearing a blue dress sitting at a table outside a café. A subjective description on the other hand might tell us that the woman's expression makes us feel sad, that perhaps she feels lonely, maybe shunned.

Whilst the latter approach may result in something which is more poetic, this style of language has limited value in a safety argument. The strength of a safety argument comes from its focus on objectivity rather than subjectivity. It should persuade the reader on the basis that the argument is compelling, credible and trustworthy. An objective approach will set out the facts, elaborate the detail, point out any apparent inconsistencies or conflict and draw conclusions from that data. A subjective approach on the other hand would describe how an individual feels about the information, what from their experience they believe about it and what impact that might have. The key difference between these approaches is that a subjective view is built on opinion and opinion can involve bias. By approaching CRM from a strictly objective viewpoint, bias is not only removed from the argument but it is seen to be removed.

So is there any room for opinion in CRM? Ultimately operating in CRM requires us to make judgements. In the ideal world we would make those judgements entirely on the basis of hard, indisputable facts. But the real world of safety, just like the law, is generally less black and white. There comes a point where we are forced to draw conclusions from less than perfect information. We can manage this by ensuring that where we are forced to make use of opinion that we clearly set out the steps and thought-processes through which we have travelled. A clear rationale provides context and demonstrates that bias has not had influence in reaching a particular conclusion.

18.4.2 Objective Writing Style

Objectivity is often conveyed through the writing style used to document a safety argument. If one is not careful it is quite easy to communicate a completely objective assessment in a manner that appears to be, at least partly, subjective.

Take a couple of examples:

Example 18.1
"When reviewing the software, I found a bug. If a patient has more than three allergies, the screen only shows the first few. You have to click a button to see the rest. What's more, the severity isn't shown. The first three might be minor reactions whilst the fourth, the one that is not shown, could be catastophic! This is unsafe, a massive clinical risk and totally unacceptable. I need to be able to see all allergies and not just a few."

Example 18.2
"On reviewing the user interface, the display of known allergic substances was found to be sub-optimal. The list of allergies is constrained to three items only. Which three are displayed is dependent on the order in which they were entered by the user rather than the clinical severity. A clinician could incorrectly assume that a patient only suffered from three, low severity allergies when actually this was not the case. The clinical risk is assessed as; Severity: Moderate, Likelihood of harm: Occasional, Clinical Risk: Medium. It will be necessary to investigate this with the software supplier to mitigate the clinical risk further."

Example 18.1 is written from the author's perspective. Whilst his/her frustration is palpable the introduction of subjectivity reduces its credibility. Interestingly Example 18.2 evaluates the clinical risk at a lower level (Medium rather than "massive") and yet its gravitas arguably enhances its ability to convince that this is a serious issue.

Some writing style pointers are offered here:

- Words such as 'I' and 'you' have no place in safety case writing style. Using these words immediately portrays the text as being personal and subject to opinion. For example, "The software was reviewed" rather than "I reviewed the software".
- Avoid unnecessary qualifiers such as 'very', 'really' and 'totally' and instead describe the facts. For example, "The screen took more than ten seconds to load" rather than "The screen loaded very slowly indeed."

- Avoid exclamation marks, words written in capitals, bold, etc. to add emphasis to the meaning, For example, "The screen loaded **VERY** slowly!"
- Never describe something as being 'safe' or 'unsafe'. These terms are meaningless in this context as after all, no system is ever truly 'safe'. The clinical risk associated with a particular feature or piece of functionality is either acceptable or unacceptable but not safe or unsafe.
- Communicate the degree of clinical risk using consistent and agreed language as set out in the SMS (e.g. Low, Medium, High). Other subjective terms like "huge clinical risk" are meaningless.
- Describe the impact of the issue accurately, unambiguously, unemotionally and within the agreed scope of CRM. For example, "This could prevent a clinician from being able to access information which he/she needs to support clinical care" rather than "Our doctors will need to work harder and longer to find the right information."

Basically the more measured and factual the language the more credible and objective it will appear to the reader. The more extreme and exaggerated the language, the more subjective and less believable it will be.

18.5 Summary

- The safety case is only likely to be compelling if it is structured and logical. The SMS should contain a standard template for the safety case.
- The safety case should bring out the salient parts of the safety argument using narrative and description rather than be a rehash of the entire hazard register.
- Key controls need to be emphasized particularly where they are required to be implemented across organisational boundaries.
- The safety case which concludes that the risk is unacceptable is a useful report but needs to be carefully managed.
- Safety cases need to be staged through the product lifecycle and updates need to be planned for.
- Objectivity is the order of the day in writing the safety case content. The more measured and factual the writing style, the greater its impact.

References

1. International Organization for Standardization. EN ISO 14971:2012. Medical devices. Application of risk management to medical devices. 2012.
2. Health and Social Care Information Centre. ISB0160. Clinical risk management: its application in the deployment and use of health IT. UK Information Standard Board for Health and Social Care. 2013.

Chapter 19
Handling Faults in Live Service

19.1 Fault Reporting

No system is error free and detecting system faults during live operation is an entirely predictable event. Healthcare organisations therefore need to ensure that the necessary processes are in place to support users when a fault is detected. Similarly, manufacturers need the infrastructure to respond to reported faults, prioritise them accordingly and institute a resolution. It is in this prioritisation phase where CRM plays an important role. Standards such as the UK's ISB 0129/0160 specifically require organisations to undertake CRM activities on reported faults [1, 2] and Medical Device manufacturers are likewise obligated to carry out post-marketing surveillance on their products.

Note that the term 'fault' in this text is used somewhat loosely. Terms such as problem, issue, defect, incident, etc. all have similar but slightly different contextual meanings. Here the intention is to describe an entity which, during live service, has or could compromise a user's ability to achieve their goals. Whether or not this results in actual harm, is due to technical or human factors, code or configuration is not distinguished in terms of the processes needed to detect and manage it.

A fault log is critical to the effective management of live issues. At a minimum this should capture the nature of the problem or incident, preconditions, timings and the identity of the reporter. Of course the log is only effective if it is supported with the appropriate business processes to manage incoming calls, data capture, investigation, routing, escalation, etc. Most organisations tackle this by implementing a service desk staffed by trained personnel working to agreed processes and policies.

Clearly not all faults share the same level of urgency and healthcare organisations need to establish with the manufacturer the methodology that will be adopted to agree and communicate the assigned priority. At the extremes the priority will determine whether a fault is fixed the same day or be added to a 'wish list' of future functionality which may or may not be delivered in the lifetime of the contract.

A. Stavert-Dobson, *Health Information Systems: Managing Clinical Risk*,
Health Informatics, DOI 10.1007/978-3-319-26612-1_19

Where manufacturers have contractually agreed resolution times, there exists a clear chain of influence from the CRM assessment to the commercial pressure to resolve it.

Such is the importance of fault priority setting that it should be ascertained precisely (and usually contractually) which party will be responsible for ultimately setting the priority level. There are pros and cons to the manufacturer and healthcare organisation undertaking this important task. Healthcare organisations are in a good position to understand the nature and scale of the fault whereas the manufacturer is able to see the bigger picture and balance priorities across multiple customers. Whatever the mechanism implemented, manufacturers and healthcare organisations are well advised to maintain a close relationship when it comes to fault reporting. An honest and respectful dialog goes a long way to oil the operational cogs.

Many factors need to be taken into consideration when assessing the priority of a reported fault. As well as the introduction of clinical risk these are also likely to include:

- Security risk
- Information governance impact
- Financial impact
- Usability issues
- Effect on reporting and statistics
- Reputational impact

Thus, there are many factors which could justifiably trigger escalation and not just a potential impact on clinical risk. A fault can have little or no clinical risk at all and yet warrant the highest level of resolution priority on the basis that it has significant security, statutory reporting or financial consequences.

Nevertheless, clinical risk is clearly a major contributor to the potential impact of a fault. Irrespective of other possible outcomes, an adverse effect on patient safety is likely to promote the priority level significantly. Healthcare organisations and manufacturers therefore need to assess reported faults to determine whether or not they are safety-related and, where they are, determine the level of clinical risk. This can be a challenge where large numbers of faults are reported and it may be necessary for filters to be put in place such that service desk staff can escalate faults of greatest concern for a formal risk assessment.

Healthcare organisations and manufacturers should share a moral obligation to assess the clinical risk of faults accurately and without undue bias from extraneous pressures. Allowing commercial, political and sometimes personal factors to influence our risk assessment can quickly lead to a cultural over or under prioritisation of faults.

For example, some manufacturers occasionally attend to faults at a rate which is unsatisfactory to their customers. This can lead healthcare organisations to overestimate risk in an attempt to bring their issues to the front of the queue. Other healthcare organisations competing for fixes have no option but to over-prioritise their issues and a highly toxic situation results. The manufacturer is no longer able to realistically see the wood for the trees once every fault is assigned maximum priority. To avoid this healthcare organisations have a responsibility to assess faults objectively and only cite a compromise of clinical safety when this is genuinely the case. Similarly, manufacturers have an obligation to play their part and fix faults in a timely manner.

19.2 Assessing Faults in Live Service

Whilst most faults will not warrant a fully documented safety analysis, it is difficult to convey the importance of undertaking a formal assessment for significant safety-related incidents. Managing a serious fault in the live service can consume a significant amount of resource, often at a time which is unplanned and when personnel are engaged in other day-to-day activities. Creating a concisely documented safety assessment creates a single point of truth which sets out a rationale for the safety concern. This arms stakeholders with the information they need to prioritise and allocate resources to manage the fault accordingly.

A succinct and evidence-based account of the problem quickly becomes a mutual focus point of an investigation. Without this the concerns of the CRM team can become nothing more than a lone voice in a sea of fast-moving communications. Note that an issue assessment can be written at any time once the nature of the fault is at least partly understood. Should one wait until a fault is resolved before setting out the risk, the horse has bolted and the opportunity for escalation has essentially passed.

Assessing a safety-related fault is similar to documenting an argument in the safety case – only this time the trigger for the hazard is real rather than hypothetical. Also at time of writing the analysis, information is likely to be incomplete and the risk incompletely (if at all) mitigated. There are many ways of documenting the issue but a sensible approach is to:

1. Define the problem
2. Establish the scale
3. Describe the potential safety impact and set out the controls
4. Evaluate the risk and make recommendations

19.2.1 Describe the Problem

One might think that a description of the problem itself a largely redundant component of the issue assessment. After all for an issue to make it as far as requiring a documented assessment, surely it is already understood well enough? In practice one finds that concisely articulating the problem is one of the most valuable parts of the assessment. In particular, setting out the description as a series of logical steps forces one to tease out the questions necessary to evaluate the risk.

Describing the problem accurately also defines the basis on which the risk evaluation will subsequently be made. Importantly soliciting stakeholder's feedback on the issue assessment will reveal whether the author's understanding of the problem is truly correct and therefore validates (or otherwise) the risk evaluation. A concise problem description requires a particular type of writing style and, interestingly, not one which we might reasonably expect.

In an age of email communication we naturally adopt forms of writing which are most effective for the medium. The modern style is often journalistic in nature – we acknowledge that those around us have little time and therefore construct our communications with an eye-catching hook at the beginning gradually fleshing out levels of detail as the narrative progresses. Those in the media have long used this technique to grab attention and let readers discontinue their appraisal at any point whilst still absorbing the most salient points in the text. What is interesting is that this technique produces poor results when used to author problem descriptions.

Let's take an example:

> Patient appointments are going missing because the transfer screen does not work correctly when the new date is in the same week. When the user presses the transfer button he/she can no longer see the patient. The functionality worked fine before the recent upgrade. Normally patient appointments can be rescheduled without any difficulty.

The author here has got straight to the point, headlining the problem description with the fact that appointments are going missing. This certainly grabs the attention but the difficulty comes when one tries to comprehend what is really going on here at a detailed level in a vacuum of context. One has to read through the information a number of times and gradually piece the picture together and constructing assumptions to provide the narrative glue. A better approach is to think how we might paint this not as a news article but instead as a short story. Consider the structure of a children's fairy story – one might set the scene, introduce the characters, their motivation and intentions. Gradually the narrative will develop the story and characters leading to a conclusion. Whilst not all problem descriptions will end with happily ever after, we can learn a lot from adopting this literary approach in our problem descriptions:

1. What clinical business process is the system supporting? What does this system/function do under normal conditions?
2. Where is the activity taking place? Which users are affected? What were they trying to do?
3. How is the system behaving now as a result of the fault? When did that change occur?
4. Under what conditions does is the problem triggered?

Applying this style to the previous example we might obtain the following:

> Administrative staff need to be able to reschedule patient appointments in the 'clinic management' screen. A problem has been identified in this area since a recent software update was deployed. When a user attempts to reschedule an appointment a fault is observed - the original appointment is deleted but the new booking is not created. The situation only occurs when the new date is in the same week as the original date.

Note that this version of events is not only clearer but isn't significantly longer and takes no real extra effort to author. Importantly it contains much more of the information one needs to evaluate the clinical risk and yet it remains concise.

In assessing issues, context is everything. The fact that 'the red text should be blue' or 'the button shouldn't be greyed out' tells us no more about the story than 'grandma was eaten by the wolf'. It is the user's intentions and the impact of the fault that causes harm not the fact that the screen won't scroll. Similarly one should always ensure that sufficient information is provided to give a view on how likely the issue will be detected by users and whether there is any potential to mislead a clinician.

19.2.2 Establish the Scale

Understanding an issue requires us to appreciate both its nature and scale. The latter can easily be forgotten, we observe a fault during an interaction with the system and report it but without any view on the size of the problem it can be difficult or impossible to assess. Gross faults which occur consistently across the system are usually found during testing and are often fixed long before go-live. In service faults are often more subtle occurring intermittently or only affecting certain users. It is important to establish at an early phase of investigating a fault what the pre-conditions are to it being triggered and from this the scale of the problem can be estimated. As a minimum one is likely to need to know:

- The number of users affected
- The number of workstations affected
- The number of patients affected
- The regularity with which the issue occurs

Determining the scale of a problem requires time, experimentation and a little detective work. The individual reporting the issue may not have the skills or patience to investigate and the task may need to be escalated to administrators supporting the system. Either way, conclusions regarding the risk associated with the problem should not be drawn until at least some attempt has been made understand the scale of the issue.

Our arch enemies in estimating scale are intermittency and detectability. Faults which only occur from time to time or those which are hard to recognise make it difficult or impossible to accurately assess scale. In some cases one has to rely on anecdote alone in which case assumptions need to be made in setting out the clinical risk. An understanding of the scale is also likely to change as the issue is investigated. Once the root cause is found it may become apparent that the problem will only be manifest when many unlikely factors align. In contrast over time more instances of the issue may be reported perhaps across different care settings and user roles. This provides us with an opportunity to revisit the assessment and, in particular, re-assess the likelihood component of the established clinical risk.

19.2.3 Outline the Potential Impact and Set Out the Controls

As with all hazard assessment work, an attempt should be made to set out the potential impact of the issue on the patient in order to set the severity component of the clinical risk. The impact should be credible and realistic in the clinical environment under consideration. The focus should be on harm not inconvenience to the user, reputation of the system or failure to realise benefits.

One might argue that once a safety-related issue has been identified it is too late to consider potential controls but this is far from the case. Just because a hazardous scenario may result from the fault it is by no means a foregone conclusion that actual harm will occur. There will usually be a human operator between the system and the patient who can work around the issue or take remedial action. In some cases, the provision of practical advice to users from a system expert may alleviate much of the real world risk until the issue is fixed. Important mitigations of this nature should be documented in the issue assessment and provide one means by which the advice may be propagated.

19.2.4 Evaluate the Risk and Make Recommendations

The risk associated with the issue should be estimated and evaluated in the usual way as outlined in previous chapters. Attention should be paid to the key factors mentioned earlier which influence risk; clinical dependency, detectability, presence of workarounds, etc. (see Sect. 14.1). There may however be some thought required when it comes to ascertaining the likelihood component of risk. The question is this, if an issue has already occurred is the likelihood not 100 %? This seems a reasonable chain of logic to begin with but it suffers from a couple of complexities.

1. Probability does not necessarily change just because an event occurs. The likelihood of rolling a six on a dice is one in six. Just because we happen to roll a six on our next turn does not change that probability.
2. Likelihood in the context of clinical risk is about the probability of harm occurring and not that a fault may create a hazardous situation. Thus even if the system consistently manifests a defect it is by no means guaranteed at all that actual harm will occur (note that consistent faults often have lower clinical risk as users become familiar with them).

Thus we should take a real-world view, given the nature and scale of the problem, how likely the issue is to cause harm of the stipulated severity on an on-going basis.

Recommendations set out in the issue assessment should provide a view on the appropriate level of priority to resolve the problem. Direction can also be offered on the priority that formulating and communicating workarounds takes. This enables stakeholders and senior managers to gauge the resources that should be allocated to the problem. Also, it can be helpful to document what additional measures should be put in place in the event that this issue is not fixed within a defined time period.

19.3 Escalating a Fault to the Manufacturer

Reported issues which cannot be resolved locally warrant escalation to the manufacturer. This critical interaction is an opportunity to ensure that the manufacturer addresses the concern with an appropriate level of priority; too low and the hazard persists for longer than is reasonable, too high and valuable resources are diverted inappropriately. Healthcare organisations can greatly assist manufacturers in providing the salient information that is needed to:

(a) Help them understand the nature and scale of the fault
(b) Understand the healthcare organisation's assessment of the clinical risk

Documenting a clinical risk assessment takes time and effort. But, where a healthcare organisation wishes a fix to be prioritised on the basis of its impact on clinical safety, it seems reasonable for that request to be backed up by an objective analysis. Manufacturers are often more than willing to prioritise serious concerns but support staff who are often non-clinical need the right information to justify escalation. Where the issue is in relation to a user interface component, anonymised screenshots are vital on the basis that a picture paints a thousand words and that safety issues are often very subtle.

19.4 Challenges to the Assessment

Occasionally stakeholders disagree on the conclusions set out in the issue assessment, often the conflict is across organisational or departmental boundaries. Stakeholders may feel that the estimated degree of clinical risk is unrealistic or its derivation illogical. Providing individuals with an opportunity to challenge the assessment rationale is not a weakness but a strength of the CRM system. By formally documenting the current understanding of a particular problem it is possible to lay out the established facts and assumptions in order to solicit opinion and agreement. Where parties disagree, this provides an opportunity to examine the basis of the divergence in opinion and potentially revisit the analysis.

Challenges to the assessment often take the form of a disagreement on:

- The nature of the fault itself
- The impact that the fault has on the patient
- The controls which exist, or
- The assessment methodology employed

Occasionally manufacturers fail to fully understand the clinical implications of a fault, especially where there is a lack of clinical input into their fault logging and investigation process. Healthcare organisations may need to help the manufacturer to understand precisely how the system is used, why it is relied upon in particular circumstances and how the fault results in changes in user behaviour. Similarly, healthcare organisations sometimes don't fully understand the technical intricacies of the product or the full functionality and configuration options provided. This knowledge might expose a readily available workaround which could change the view on the clinical risk.

Where a stakeholder criticises the method used to derive the assessment, this may need to be tackled differently. Here the focus should be on reference to applicable standards, best-practice and agreed process. A solid and documented CRM system with historical good standing is a robust starting point for demonstrating that the assessment approach is valid and sound. Often when the methodology is criticised it is simply the chain of logical thought which has failed to be articulated fully rather than a genuine fault in the process.

In summary, a succinctly documented issue assessment is not a means of apportioning blame but rather a tool to stimulate discussion, prioritisation, mutual understanding and ultimately timely resolution.

19.5 Managing Serious Faults and Incidents

Just occasionally a very serious issue may be found which either results in actual harm or which precipitates a hazardous scenario which cannot be tolerated. These situations need to be very carefully managed with close co-operation of all stakeholders. The key to an effective approach is planning. In formulating the CRM system one should draw up a simple series of steps to be undertaken when faced with a serious incident. In particular thought should be given as to which stakeholders should be immediately engaged and where their contact details will be maintained. Stakeholders might include representation from the organisation's legal and communications departments as well as senior management. In the melee of a serious incident one needs to quickly separate fact from here-say. The rapidity with which a clinical risk assessment is required is not a license to resort to anecdote over observable evidence.

The first priority of course is to the patient or patients who may have experienced harm. Direct communication with the patient should be undertaken by senior clinicians briefed on the situation. Even at this stage it may be possible to employ reactive controls to minimise or contain the harm experienced. Ideally in parallel there should begin a thorough investigation of the issue the initial focus being placed on a 'make safe' objective. Whilst the underlying cause may not be known it is sensible in the first instance to consider what can be done to mitigate any on-going and immediate threat. For example, suppose a system begins to spuriously report incorrect drug dosages. One might immediately notify all prescribers and drug administrators to check all prescriptions and employ extreme care in trusting dosages indicated by the system. Whilst this does not represent any kind of solution to the problem the extra vigilance may improve detectability and contain the problem for the next few minutes or hours whilst a more thorough assessment takes place.

Once the immediate danger is handled, the focus should be on establishing precisely the nature and scale of the problem with a view to determining the root cause. Once that is known the true scale can be ascertained and a more informed view on the clinical risk taken.

Occasionally one is faced with a difficult dilemma when the only realistic 'make safe' option is to discontinue use of the system (or part of the system) until the issue is fixed. This decision should not be undertaken lightly as this may well introduce risk in itself. In this case one is essentially exchanging the risk of the fault with that of system unavailability. Making an informed decision is significantly easier when the risk of unavailability has been proactively assessed and documented (see Chapter 7). Thought should also be given as to how technically this might be achieved; simply requesting that users discontinue operating the system without any actual barriers being put in place might not be wholly effective.

19.6 Implications for the Safety Case

Much of the safety case will be authored prior to system go-live but we have already discussed how the hazard register and safety case should be living documents. At the time of authoring the pre-go-live safety case, assumptions and estimations will have been made regarding the clinical risk. Often it is only during live operation that these can be checked. A number of safety standards require that the controls identified in the safety case should be validated for their presence and effectiveness. Validating presence before system go-live is often straight forward but testing effectiveness less so. For many controls, the only way to test effectiveness is to monitor operation of the system in live service.

Organisations may choose to actively monitor specific controls (where the level of risk justifies this) or to passively collect reported service incidents and test these against any assumptions made in the safety case. Either way it is likely that during live operation one will determine that there are some controls which prove to be less effective than others and a judgement call needs to be made on whether a reassessment of the clinical risk needs to be made for individual hazards. Where controls are found to be ineffective, it is generally the likelihood component of the risk calculation which may need to be revisited. In some cases, new controls will need to be implemented to manage the risk. It is only by the careful review of reported faults and examination of the hazard register that one is able to determine that there is a case for expending cost and effort in developing additional mitigation.

Similarly, live service may reveal new hazards or causes. It is not uncommon for complex systems to experience failure modes which couldn't possibly have been foreseen during design. In some cases operation of the system may reveal that the product is used in such a way that the hazards have a very different impact on the patient to what was initially conceived. Capturing and preserving this knowledge in the hazard register is vital for informing system changes and developments in the future as well as disseminating learning across organisational boundaries.

Manufacturers and healthcare organisations will also need to make a judgement call on whether new information coming to light warrants not only an update of the hazard register but a re-issuing of the safety case as well. It is useful to document in the SMS the criteria under which the safety case would be updated and re-issued during live service. This might include for example:

- A significant increase in the clinical risk
- A need for new controls which require implementation across organisational boundaries
- Identification of significant new hazards, causes or impacts

19.7 Summary

- All systems develop faults at some point in service and in HIT it is almost inevitable that some will be related to safety.
- Both healthcare organisations and manufacturers need to actively support the reporting of incidents by users such that they can be addressed in a timely manner.
- Not all faults impact safety to the same extent so a CRM assessment will be required. The degree of clinical risk should be one of the factors which influence fix priority.
- Objectively documenting a CRM assessment for a significant fault is a powerful tool in gaining stakeholder buy-in and effecting its resolution. This is particularly the case when a healthcare organisations needs to escalate a fault to the manufacturer.
- Faults provide a rich scheme of knowledge and the learnings may reveal new information which needs to be incorporated into the safety case and the risk re-evaluated.

References

1. Health and Social Care Information Centre. ISB0129. Clinical risk management: its application in the manufacture of Health IT Systems Version 2. UK Information Standard Board for Health and Social Care; London. 2013.
2. Health and Social Care Information Centre. ISB0160. Clinical risk management: its application in the deployment and use of health IT. UK Information Standard Board for Health and Social Care; London. 2013.

Chapter 20
Maintaining the Safety Case

It is unfortunate that many a safety case is constructed before product launch only for it to be filed away in the annals of history after go-live. The safety case is a key document for the lifetime of the product but its content shouldn't be static. There is much to be learnt from operating a system in a live environment and the world changes in ways that could impact the validity of the arguments set out in the safety case. It is the nature of the software industry that products evolve, new functionality is introduced and old methods are deprecated. Clinical processes change too and in a short space of time there can be extensive divergence between the original safety case and the product as it stands today.

In the event that a serious incident was to occur and the system put in the frame, investigators will quickly turn to the safety case. An out of date safety case is a prime target for criticism and failing to maintain such an important report brings into question an organisation's entire safety culture. Organisations need to develop trigger points which automatically and habitually prompt a review of the safety case and elicit a possible revision.

For most projects it is good practice to re-visit a safety case on a regular basis and make a critical evaluation to determine whether or not it is still valid – an annual review is common in many industries. The review strategy should be set out in the SMS along with other 'in service' safety activities.

20.1 Managing Change

Most Health IT services change over time, whether it is the underlying technology, system code, configuration, its implementation or any number of other factors. In fact one of the properties of a software system is the apparent simplicity by which they can updated compared with the practicalities of modifying physical devices. This flexibility brings significant benefits.

A. Stavert-Dobson, *Health Information Systems: Managing Clinical Risk*, Health Informatics, DOI 10.1007/978-3-319-26612-1_20

But that ability to adapt the system to changing requirements over time is both a blessing and a curse in terms of risk management. Should a defect be detected it may be possible to implement a fix through a patch, script or hotfix without the need for any system downtime. However, such a moving target creates a problem when it comes to assurance. Suppose the patch fixes one problem but introduces another; suppose the operator in applying the patch inadvertently overwrites a file or incorrectly modifies the configuration? Even the smallest of technical changes has the potential to adversely impact the system in some way if the process is not carefully controlled and thought through.

Similarly working practices can drift away from those the system was originally designed to support and when this occurs the likelihood of unsafe outcomes increases [1]. Often these changes are subtle, organic and undocumented but can amount to extensive deviation given sufficient passage of time.

The ECRI Institute's Top 10 Health Technology Hazards for 2014 places neglecting change management in networked devices and systems at number seven [2]. The report cites several examples of incidents including an update to an electronic health record which resulted in the loss of date and time data from radiology reports.

20.1.1 Sources of Change

Many types of change can take place both in the technology and the way it is implemented. Being aware of these modalities can help detect where and when change might occur.

Examples of changes to the HIT system:

- The software code
- System configuration
- System architecture
- Server hardware
- Server or client operating system
- Network infrastructure
- Underlying database or its configuration
- Housekeeping procedures
- Client platform or display size
- Failover procedures

Examples of change in how a HIT system is used:

- Scope of functionality in operation
- Clinical processes supported by the system
- Range of care settings
- Operating policies and processes
- The type and number of users
- Local configuration
- Business continuity procedures

With such a large and diverse number of factors to consider, it is hardly surprising that change is common and perhaps more common than one would care to realise. Uncontrolled change is a threat to the integrity and validity of the safety case. This is where a sound change management process comes into its own. By having a centrally managed repository of planned changes, modifications to the service can be detected, impacted and implemented with minimal disruption to clinical business.

20.1.2 Detecting Change

For many organisations the greatest challenge can be detecting change at all. Increasingly healthcare organisations are adopting software as a service where the system is remotely hosted and managed – local staff have little or no hands-on responsibilities. In these cases the manufacturer is technically able to make modifications to the system at will with little or no notice to its customers. Healthcare organisations can be left wondering why system behaviour has changed overnight without warning and with no opportunity to adapt local processes or re-train staff. This failure of change management can pose a significant threat to safety and handling this situation is largely down to ensuring that change processes are agreed contractually. In this way manufacturers become legally obliged to provide their customers with details of the change in a timely manner.

Most manufacturers issue their customers with at least some documentation outlining proposed changes. Healthcare organisations need to factor in the time required to analyse these materials and determine any potential impact on the safety case. Release notes contain varying levels of detail and a view must be taken on whether this is sufficient to understand the change well enough to meaningfully adapt. For manufacturers who carry out their own safety work and communicate this to customers in good time, the task of managing change is greatly simplified.

Even within organisations, change can be difficult to detect. For example, a local configuration change could fail to be communicated to all users and/or the training team such that their materials are out of date and no longer valid. Changes at a local level therefore need to be subject to exactly the same level of analytical rigour as those undertaken by external parties. All in all, without a process in place to detect change, stakeholders will fall at the first hurdle and the modification will be implemented before individuals have any opportunity to consider the consequences.

20.1.3 Impacting Change

Once the scope of a change is known, the next task is to understand the effect of that change on the system and any implications for how it should be operated. Changes to software systems often exhibit a characteristic known as the ripple effect. Modification of one part of a system can impact another which in turn has baring on yet another. Thus the end state can be difficult to predict; where there is unpredictability, this can compromise the arguments presented in the safety case.

Successfully analysing the effects of a change depend largely on the extent to which the relationship between its components are sufficiently defined. Well documented systems often depict traceability or dependency structures. Using this one is able to follow a chain of logic from the element being changed to those others potentially affected. This information makes it possible to model the likely behaviour of the system after the change. For systems with less documentary rigour it may be necessary for technical staff to manually unpick the code or configuration and determine from first principles where dependencies lay. For sure, a strategy of "let's change it and see what happens" in a live environment is not a realistic approach in safety critical systems.

The process of impacting change is another where a multidisciplinary approach works well and the notion of a 'change board' can provide the necessary governance to ensure the right stakeholders are engaged. Representative from the CRM team should be key attendees at these sessions and take note of all parties which could be affected by a change.

Even small changes can necessitate staff briefings, updates to training material, changes to configuration, etc. It is those activities which need to occur in order to maintain safe operation which is of particular interest from a CRM perspective. One needs to consider whether the change introduces new hazards or whether existing controls could now be less effective. The techniques used in the analysis should be no different from those applied to new functionality as described elsewhere in this book. Finally the clinical risk needs to be re-assessed and determined whether or not the new risk profile is acceptable. Of course this should be articulated through an update to the hazard register and safety case.

Interestingly the clinical risk associated with system change is often far greater than the risk associated with the system when it first goes live. When new functionality is introduced for the first time users are naturally cautious and unexpected system behaviour which could compromise safety is readily reported by alert users. If the risk of operating the newly installed service turns out to be unacceptable, the chances are that old processes are still supported and clinicians can return to using paper, the telephone or other means of maintaining business continuity. But when a change suddenly causes a tried and tested system to fail, users may find that the old ways are simply no longer an option and a hazardous situation begins to unfold.

20.1.4 Implementing Change

When it comes to implementing a technical change in HIT systems thought needs to be given to the potential scenarios which could result immediately or shortly afterwards. A number of situations where change is unsuccessful are possible (Table 20.1):

In implementing a technical change organisations need to have a plan in place in case the deployment is unsuccessful or the resulting system is left in a state of intolerable risk. This will often involve a back-out strategy where the change is reversed

Table 20.1 Potential changes in clinical risk after implementing a change

Scenario	Resulting clinical risk
The change itself fails but otherwise leaves the system unaffected	Unchanged which could be an issue if the modification was to address an existing intolerable hazard
The newly introduced functionality turns out to be associated with new unexpected hazards	Depends on the nature of the new hazards
The change adversely impacts other dependencies which were not previously foreseen	Depends on which components are affected and to what extent
The change, or the process of implementing the change renders the system unavailable or unusable	The same as that established for general system unavailability

and the system restored to its pre-change state. But how and when the change is backed out can also have safety implications. For example, suppose a change is implemented over a weekend but serious safety-related problems are not detected until Tuesday. What impact will backing out have on Monday's data? Thus, significant changes should be associated with a formal transition plan setting out the strategy and be subject to a safety review.

20.1.5 Removing Functionality

An interesting situation can arise when there is a need to remove functionality from live service or de-scope it close to go-live of a new service. There may be a number of reasons why this might be desirable, for example:

- The functionality is associated with unacceptable clinical risk and a strategic decision is made to eliminate this by withdrawing it from service.
- The functionality is subject to specific licensing and its continued operation is no long commercially viable
- The functionality has been replaced by an alternative approach and is no longer required

In these circumstances, the key question is whether the functionality being removed is acting as a control and could the action therefore result in an increase in clinical risk? Similarly, would removing the functionality prevent access to any key data already captured? For example, suppose a system has functionality to record a patient's social history as free text. A change is proposed to replace this with a more structured data capture form. It would therefore seem reasonable to deprecate the original functionality, after all having two disparate places to record the same information would pose a hazard in itself. However, without careful design the information already captured as free text could become orphaned and inaccessible. A more strategic and safer approach might be to render the old functionality read-only or to provide a data migration of the old data into the newer forms.

20.2 Evolution of the Safety Case

20.2.1 Changes in Risk Acceptability

The implementation of change is not the only trigger for updating the safety case. From time to time it is not the system or its implementation which changes but the world around it. In general there is a trend for society to become more risk averse and as such governments, regulators and purchasers need to find opportunities to raise standards of safety without stifling innovation. Changes in legislation and revision of standards can mean that a safety case and/or SMS which once met the necessary requirements suddenly has shortcomings. Remember that the safety case needs to be valid for the entire life of the product so just because a standard or regulation was once complied with it cannot be inferred that the product will always remain compliant. A change to a standard of course doesn't mean that the existing product is any less safe but the argument and evidence may need to be developed further or be set out in a particular way for the safety case to meet the new requirements.

Another important consideration is the continued validity of the ALARP argument. As discussed in Sect. 3.1 ALARP is essentially a balance between risk and the resources required to mitigate it further. These elements are not static but represent a movable feast which need to be revisited over time. Risk acceptability is dictated by society's view of what is tolerable in a given set of circumstances. Rightly or wrongly these values are strongly influenced by the media and events in living memory. Similarly the time, effort and cost of implementing a particular safety feature can change significantly over time. Suppose for example that one control option is the hourly off-site backup of the entire system database. At the point of system go-live the network bandwidth and storage capacity required might be technically and financially prohibitive and therefore its rejection as a control be justified. Five years later when perhaps the technology is much more affordable this argument may no longer stack up and the options may need to be revisited.

20.2.2 Learning from Operational Experience

The safety case developed prior to go-live inevitably contains assumptions about the live operation of the product. In complex systems it is hardly surprising that some assumptions turn out to be invalid whilst other previously unforeseen characteristics come to light. These learnings don't necessarily represent a failure of the safety case but rather a basis for its next stage of evolution. For example, suppose prior to go-live, a system undergoes volume and performance testing in order to determine an appropriate hardware specification. It may subsequently be found that the transaction mix of the service in the real-world places a greater demand on the system than was expected. The original clinical risk associated with poor performance may have been evaluated as low given the evidence of the test output but in hindsight this may need to be revised until further controls can be put in place.

Learnings capable of affecting the validity of the safety can come from many different sources. For example:

- Reported faults
- Logged incidents and near misses
- Interviews with users (both new and experienced)
- Availability and performance data
- The content of error logs
- Experience of system trainers

One important source of data which is often forgotten in HIT is the experience of other organisations using the same system. Compared to other industries, healthcare organisations seem surprisingly reluctant to learn from their colleagues elsewhere. A significant safety-related fault or limitation of the system spotted by one organisation is likely to be relevant to another or at least could be in the future. Baring in mind that some manufacturers have hundreds or even thousands of implementations, this rich source of data (if one is able to access and formulate it) can be vital in challenging assumptions made in the safety case.

Another pointer is to consider the experience of other healthcare organisations using different systems but in a similar context or care setting. Many hazards are not unique to individual systems but are rather shared amongst applications supporting the same clinical business processes. The range of control options employed by different manufacturers offers a real-world experiment in effectiveness. The sharing of this data in the HIT industry is still not commonplace and much can be learnt from the transparency of aerospace manufacturers. Nevertheless even at an anecdotal level, one is wise to take heed of important observations at other institutions and test them against the assumptions set out in the latest iteration of the safety case.

Finally remember that whilst operational experience can sometimes challenge the claims in the safety case, one would hope that it would also reinforce it. Where a control strategy has found to be particularly effective then enriching the safety case with this real-world evidence builds a very strong argument indeed. This is especially the case for novel hazards or experimental controls where, at the point of go-live, effectiveness to some extent is assumed but cannot be completely evidenced. Taking the time to update the safety case preserves this knowledge for future iterations of the product and future generations of personnel.

20.3 Decommissioning HIT Systems

Every HIT system has its day and it is inevitable that at some point a system will be decommissioned and replaced by an alternative. One might assume that it is at this point that the safety case is effectively switched off and replaced by another covering the demands of the replacement service. However it should be noted that the decommissioning activity itself is generally a safety-related task. In fact decommissioning comes with some interesting characteristics of its own to the extent that the process is likely to warrant a safety assessment and quite possibly its own safety case.

Many decommissioning projects involve a data migration from one system to another. A data migration is the relocation of data from one database or application to another. The activity is different from the type of continuous data feed one would see with an interface between live systems and generally takes the form of a one-off bulk transfer of data which completes in a discrete time window. Many challenges are faced when a data migration is planned and this is often confounded by the need for a number of organisations to work closely together. This may for example require the co-operation of two or more suppliers who are traditionally competitors and where one might be the subject of a contract coming to an end. Project managing a data migration can call for some careful diplomatic and commercial skills.

In HIT data migrations come in many forms are not just limited to decommissioning. They can be required for example when:

- Upgrading from one major version of an application to another
- Upgrading the version of a database or moving from one proprietary database to a different one
- Mergers of systems or organisations
- Relocation of systems

The migration of clinical data is nearly always a safety-related task and should be subject to a formal safety assessment. However, whereas most hazard registers describe the on-going (and indeed maturing) experience of operating a HIT system, the data migration hazard register might only apply to a single, one-off event. Once the migration has completed and the data verified, any hazards described in the hazard register will either have been triggered or will no longer be applicable.

Data migrations usually require the movement of large volumes of data. With day-to-day system operations a user is typically only able to impact one (or perhaps a few) patient records in a single transaction. The impact of any faults in these transactions is therefore relatively limited (unless they are present and undetected for a long period of time). In contrast, a data migration has the potential to adversely impact every patient in the system in a single activity. Thus a significant degree of rigour is called for in the design and execution of the task.

Data migration hazards frequently arise in the following areas:

- Inappropriate or incorrect transformation of data
- Incomplete or partial migration
- Corruption of data
- Failure to manage differences in reference data
- Mismatch of data models
- Potential for downtime during the migration
- Adverse performance impact during data extraction

Data migrations typically fall into two categories; simple migrations where the data is moved in its entirely from once place to another but essentially remains unchanged and transformations where the data undergoes some kind of modification on route to its destination. Data may be transformed in a number of different ways ranging from simple data-type changes (e.g. a whole-number field into a decimal field) to complex data re-mapping. Unsurprisingly it is often the more complex transformations which can lead to unpredictable or inconsistent results. Similarly organisations often choose to migrate only a subset of the source data into the new system. This necessitates the creation of rules and logic which can become complex both in terms of their specification and coding. Get this wrong and valuable records can easily be lost in the course of the migration.

Frequently the data models of individual systems vary making it difficult to directly translate data in one system to that in another. In most areas there are no agreed standards for the formatting of clinical data – or at least few which manufacturers fully and consistently adopt. For example, suppose the originating system has functionality to record any number of alias names for a patient (using a one to many relationship between database tables). The destination system on the other hand might limit this to (say) three alias names, perhaps with three discrete fields in a single database table. What does one do with those patients with more than three aliases (which is not uncommon)? How might this impact patient identification and safety when a patient subsequently presents using one of these aliases?

A similar situation results when data is described differently between two systems. As a simple example, consider a scenario where the originating system presents a prompt as follows:

Number of cigarettes smoked each day:

- None
- 1–5
- 6–10
- 11–20
- 21–30
- 31+

Whilst the target system provides the following options:

- None
- 1–10
- 11–25
- 26–35
- 36+

How would this data be accurately transformed? In some circumstances the safest strategy might be to deliberately exclude the data from the migration on the basis that no information is better than misleading or incorrect data. The user would need to re-capture the information afresh when the patient record is next accessed. Occasionally healthcare organisations implement imperfect but creative workarounds such as importing structured data into a free text or 'notes' field to avoid it being completely lost.

The situation can become even more complex when the existing data is of poor quality. Over time users may become less disciplined in recording certain data items or re-use fields designed for one purpose for something completely different. The transformation business rules may hold up whilst ever the originating dataset is complete and intact but when this is not the case the amount of conditional logic needed to faithfully transform the data can be intellectually overwhelming. Occasionally database queries rely on certain data items being present and when things are missing, however minor, the database engine may by default omit the entire record. For example suppose a patient record is being migrated and the query expects the patient to have a 'next of kin' recorded. Whilst this might be a reasonable assumption for 99 % of patients, should this not have been recorded it is possible that the entire clinical record for that patient could fail to migrate.

The timing of the data migration is also a crucial part of the transition plan. With large databases it can take a long period of time to migrate the data, sometimes weeks. This raises an interesting dilemma, does one shut the system down and accept the risk of imperfect business continuity measures until the new system is up and running or should the migration occur whilst the system is live? Where the system is deliberately taken offline the healthcare organisation must be clear about how it intends to support clinical care during the outage and what options are available should the migration ultimately fail.

If the system remains live during the migration at least two more potential hazards emerge:

- If live transactions are still occurring at the same time as a data extract, how will the migration ever catch up and be completed?
- Could the intense data extraction process adversely impact performance of the live environment?

In these circumstances it may be appropriate to employ dedicated data migration tools and experts in using them. These often provide a live feed of data into a new environment and functionality to throttle the data extraction process so as not to impact the live environment.

Ensuring the success of a data migration is largely down to planning the activity, developing the right business rules for any transformation and then validating the resulting data set. Developing the right strategy should involve a multidisciplinary team including technical experts, representatives from the manufacturers, clinicians and system administrators. Often an iterative approach works well especially where transformation rules are complex and are subject to extensive logic. The minutes of review sessions and ultimately a formalised transition strategy should all be referenced from the safety case. Where the quality of the source data is in doubt it may be necessary to instigate a data cleansing project prior to migration – an activity which has a significant impact on success but can be labour intensive and must be incorporated into the project timelines.

Perhaps the most important piece of evidence for the safety case is the testing and validation of the final dataset. This often involves a detailed analysis of the resulting data usually done in a test environment prior to the live data migration so there is an opportunity to tweak the strategy should defects be found.

The validation approach could involve:

- inspection of the raw data by those familiar with its structure
- sample testing against the source data
- the execution of specific scripts to return erroneous, incomplete or suspicious records
- row counts of database tables
- viewing of the migrated data in the new system as it would appear to end users
- the use of specialist data checking tools

By the time the actual cut-over takes place the strategy should be sound and only its execution require final validation. Often a subset of tests are employed at cut-over in order to strike a balance between transition time and assuring the final migration.

20.4 Summary

- The safety case will become outdated if it is not maintained during live service. At a minimum, an annual review should take place.
- Software systems are relatively easy to update but without formal change control measures in place the safety case and live product can diverge.
- Change comes in many forms and processes should take account of how modifications will be detected. It is not only the system which can change but also the environment in which it is used, the numbers of users and the extent to which it is relied upon.
- When systems are upgraded, the process of implementing the change is also a safety-related activity and might justify its own safety assessment.
- Decommissioning a system can be a complex task especially when it involves a data transformation and migration. These processes have the potential to adversely impact every record and every patient in the database.

References

1. Institute of Medicine. Health IT and patient safety: Washington, DC: National Academies Press. building safer systems for better care. 2012.
2. ECRI Institute. Health devices. ECRI Pennsylvania. Top 10 Health Technology Hazards for 2014. 2013.

Index

A
Absent functionality, 146–147
Abstract conceptualisation, 227
Acceptability, 25, 39–44, 49, 51, 52, 124, 126, 130, 158, 171, 197, 205, 265, 266, 271, 292
  stratification, 41–42
Acceptance testing, 237, 238
Active engineered controls, 223–224
Active experimentation, 227
Adverse incidents
  from health IT, 11
  reporting (*See* Incident reporting)
  types reported, 11
Affordance, 72
Agile, 165–167, 272
  manifesto, 165
ALARP. *See* As low as reasonably practicable (ALARP)
Alert blindness, 225
Allergies, 71, 88, 91, 115, 176, 188, 211, 224, 226, 257, 258, 274
Analyses of variance and correlations, 27
Annex Z, 42
An organisation with a memory, 5, 65
ANSI/IEEE 1059, 233
Apps, 18–19, 76, 92
As low as possible, 40
As low as reasonably practicable (ALARP), 40–43, 45, 49, 50, 202, 206, 225, 249, 266, 292
Audit, 44, 56, 126, 131–133, 137, 147, 152, 245
Australian incident monitoring system, 77
Automated testing, 235
Availability, 81, 101–117

B
Backlog, 166
Backup, 108, 109, 137, 292
Band waggon effect, 217
Bandwidth, 111, 113, 114
Banner bar, 88, 92
Best practice, 27, 28, 47, 63, 68, 69, 71, 77, 98, 99, 112, 152, 167, 231, 256, 262
Bias, 33, 135, 215–217, 273, 274
Black box analysis, 162
Black box testing, 235, 237
Blame, 5, 28, 43, 44, 46, 61, 63–65, 77, 128, 284
Boundary cases, 141
Brainstorming, 184, 186, 192–195, 203
Bring your own device, 92
Browsers, 92, 94, 238
Building safer systems for better care, 13, 99
Business continuity, 85, 101, 108, 114–116, 208, 290, 296
Business management systems, 14–15

C
Capture slip, 67
Care administration systems, 15–16
Carrot diagram, 41
Catastrophic failure, 107–109
Causes, 83–84
Change
  control, 192
  detecting, 289
  impacting, 289–290
  implementing, 290–291
  management, 143, 159, 287–292
  mangement, 167–168

A. Stavert-Dobson, *Health Information Systems: Managing Clinical Risk*, Health Informatics, DOI 10.1007/978-3-319-26612-1

Change (*cont.*)
removing functionality, 291–292
sources of, 288–289
CHAZOP, 199
Clinical data
miscommunication, 97–98
misidentified, 89–90
misleading, 88–100
misleading presentation, 91–94
missing, 84–88
out of context, 90–91
unable to capture new, 115
unavailability of existing, 115
Clinical dependency, 206–208, 282
Clinical judgement, 99, 115, 141, 223
Clinical management systems, 16
Clinical risk
categories, 31
definition, 5–6, 30
Clinical risk management (CRM), 158–168
policy, 126–127
training, 137–138
Clinical risk management plan (CRM plan), 155–173, 266–268, 272
assumptions and constraints, 164–165
dependencies, 161–162
managing change, 167–168
project boundaries, 159–160
stating intended purpose, 160–161
Clinical safety officer, 50, 150
Cloud, 108, 114
21 Code of federal regulations, 52
Code review and inspection, 234
Cognitive bias, 23, 215
Cognitive load, 73, 207, 250, 256
Cognitive walkthrough, 253–254
Cognitive walkthrough (CW), 253–254
Colour blindness, 74
Commercial off the shelf, 136
Committee on safety of drugs, 4
Common user interface project, 71
Competency, 129, 132, 138, 149–53
Completeness, 40, 74, 83, 132, 170, 185, 187, 190, 193, 195, 219, 234, 240, 245, 270
Complete service failure, 85
Complexity, 212–214, 224
Component failure, 103
Computerised physician order entry, 8
Concrete experience, 227
Concurrent users, 111
Confidentiality, 76, 142, 239
Confirmation bias, 215
Conformance with standards, 262
Connecting for health (CfH), 50
Constraints, 267
Contextual design, 251
Contextual inquiry, 251
Controls
classification, 220–223
engineered, 223–225
overview, 219
through administration, 222
through elimination, 220
through engineering, 221
through personal protection, 222
through substitution, 221
training orientated, 225–31
Credibly incorrect data, 210
Criticality analysis, 197
CRM. *See* Clinical risk management (CRM)
CRM plan. *See* Clinical risk management plan (CRM plan)
Curriculum vitae, 150
CW. *See* Cognitive walkthrough (CW)

D
Database housekeeping, 95
Data centre, 105, 107, 108, 114, 201
Data corruption, 95–97
Data delivery failure, 86
Data migrations, 113, 212, 242, 293–295
Data synchronisation point, 108
Decision support, 6–9, 16, 17, 68, 98, 99, 224
Decommissioning, 48, 52, 272, 293–295
Deliverables, 50, 125, 126, 130–132, 137, 150, 157, 158, 239, 241
Dependability, 81, 82, 103, 206
Dependable computing, 81
Description slip, 67
Detectability, 87, 208–211, 223, 281, 282, 285
DHCP, 104, 105
Diagnosis, 6, 12, 17, 68, 76, 90, 93, 98
Diagrams, in clinical notes, 93
Disaster recovery, 101, 108–109
Disclaimers, 13, 230
Distributed alarm systems, 53
Distributed architecture, 107
Diversity, 106
Document management, 137
Domain knowledge, 29, 34, 210, 212
Downtime, 34, 42, 85, 101–116, 288, 294
Drug administration, 16, 85, 96, 115, 188, 193, 194
Dynamic testing, 234, 235, 262

E
Efficiency, of code, 112
e-iatrogenesis, 9
e-learning, 230
Electronic prescribing, 8, 16, 42, 134, 136, 147, 178, 193
Elusive data, 87
Ergonomics, 72
Error messages, 74
ETA. *See* Event Tree Analysis (ETA)
Event Tree Analysis (ETA), 183, 187, 200
Evidence
  design features, 249–250
  historical operation, 261
  operating policies, 256–259
  testing, 240–246
  user interface evaluation, 250–251
  workshops, 262–263
Expert review, 252–253

F
Failure mode and effects analysis (FMEA), 183, 197
Failure Mode Effects and Criticality Analysis (FMCEA), 209
Failure mode effects and criticality analysis (FMCEA), 197
Failure modes, 244, 285
Failure of health it, 81–84
Fault (s)
  challenging the assessment, 283–284
  describing the problem, 279–281
  escalating, 283
  evaluating the risk, 282–283
  impact and controls, 282
  reporting of, 277–279
  safety case implications, 285–286
  scale, 281–282
  serious, 284–285
  tolerance, 101
Fault reporting, 134
Fault Tree Analysis (FTA), 183, 187, 199–201
FDA, 12, 16, 18, 52, 69, 250
FMCEA. *See* Failure Mode Effects and Criticality Analysis (FMCEA)
Food and drug administration. *See* FDA
Framing effect, 216
FTA. *See* Fault Tree Analysis (FTA)
Functional failure, 85
Functional testing, 236, 237

G
Garbage in-garbage out, 94
Global harmonization task force, 17
Grey-box testing, 236
Groupthink, 217

H
Han, 11
Harm, 3, 23, 29, 35, 141
Harmonised standards, 52
Harvard medical practice study, 4
Hazard, 28–29
Hazard and operability study (HAZOP), 183, 199
Hazardous situation, 3, 35, 76, 282, 290
Hazard register, 168–169
  causes, 176–177
  controls, 177–179
  impacts, 175–176
  informing testing, 240
  populating with content, 202
  structuring, 175–181
HAZOP. *See* Hazard and operability study (HAZOP)
Health and Safety at Work Act, 40
Health IT
  benefits, 6–7
  as a cause of risk, 9
  controlling risk, 12–13
  definition, 6
  as a mitigator of risk, 8
  types, 14–19
Heuristic evaluation, 252–253
High reliability organisations, 24
HL7, 95, 123, 262
Human error, 7, 61–64, 66–69, 71, 72, 231
Human failure, 66–69

I
ICD 10, 96
Identity confusion, 89–90
IEC 61508, 41, 48–49
IEC 62304:2006, 56
IEC 80001-1:2010, 52–53
IEC/TR 80002-1:2009, 52
IG. *See* Information governance (IG)
Illusion of communication, 97
Impact of hazards, 175–176
Inappropriate denial of access, 86
Incident reporting, 76–77, 121, 134, 285
Inductive reasoning, 170, 199

Information density, 74
Information governance (IG), 86–88, 142, 239
Information Standards Board (ISB), 50
Input error, 94
Inspection of software, 241–242
Institute of medicine, 77, 99, 233
Institute of Medicine (IOM), 5, 8
Integration testing, 237, 238
Intellectual property, 13, 71, 125, 162, 267
Interfaces and messaging, 191
International Electrotechnical Commission (IEC), 48
IOM. *See* Institute of Medicine (IOM)
ISB0129, 45, 48–51, 57, 137, 150, 157, 166, 171, 271, 277
ISB0160, 45, 48–51, 57, 137, 150, 157, 166, 171, 277
ISO 9000, 53–55
ISO 13485, 52–55
ISO 14971, 31, 36, 42, 45, 49, 50, 52, 55, 137, 150
ISO 31000, 23, 27
ISO 62304, 52
ISO 62366, 52, 55–56
ISO 62366:2007, 250
ISO/IEC 25010, 234
ISO/IEC Guide 51, 29
ISO/TR 29322:2008(E), 50
ISO/TS 25238:2007, 51
ISO/TS 29321:2008(E), 50
Issues, safety, 201–203
Iterative design, 71

J
James Reason, 5, 61, 67
Just culture, 76, 77

K
Knowledge based performance, 67
Kolb, 227

L
Laprie, 82
Lapses, 67–68, 70, 72–74, 207, 214, 256
Learning cycle, 227
Learning modalities, 230
Liability, 43
Likelihood, 30–32, 35–36, 271, 281, 282, 285
Likelihood categories, 31, 32
Load testing, 238
Local service providers, 49
Logic checkers, 234

M
Maintainability, 81
Malicious act, 142–144
Master-detail workflow, 72
Maximum likelihood estimations, 27
MDD. *See* Medical Device Directive (MDD)
Mean time to failure (MTTF), 84, 103
Medical device(s), 16–18, 52, 55, 56, 77, 250
  risk acceptability, 42–43
  software, 52
Medical Device Directive (MDD), 16, 42, 52, 54, 150
Medical device regulation, 13, 16–18, 52, 234
Medical device regulators, 54
Medication errors, 8
MEDMARX, 11, 77
Mental models, 75–76
Misleading
  communicate of data, 97–98
  context, 90–91
  data due to corruption, 95–97
  direction, 98–100
  identity, 89–90
  input, 94–95
  presentation, 91–94
Mistakes, 68, 74, 144, 207, 214
Misuse of Health IT, 144
Mobile platform, 18–19, 92
Mode error, 72
Module failure, 85
Monitoring live service, 106–107
Multidisciplinary analysis, 185

N
NASA Task Load Index, 256
National Health Service, 49
National Patient Safety Agency (NPSA), 5, 49
National Programme for IT, 49
National reporting and learning system, 77
Non-functional testing, 236, 238
Normalcy bias, 216
Notified body, 55
Novelty, 214–215, 261
NPfIT. *See* National Programme for IT
NPSA. *See* National Patient Safety Agency (NPSA)

O
Objectivity, 274–275
Observational selection bias, 216
Operating systems, 161, 201, 238

Operational experience, 75, 123, 124, 192, 213–215, 242, 258, 292, 293
Organisational boundaries, 13, 32, 34, 43, 65, 98, 127, 142, 158, 168, 172, 217, 219, 283, 286
Organisation with a memory, 260
Outage, 34, 85, 86, 101–116, 176, 296
Out-sourcing, 150

P
PACS. *See* Picture Archiving and Communication Systems (PACS)
Pair programming, 166
Passive engineered controls, 224–225
Patient portals, 145
Patient safety
  definition, 3–4
  history, 4–6
  impact of Health IT, 7–12
Patient Safety and Quality Improvement Act, 5
Patient safety organisations, 5, 77
Patient Safety Rule, 5
Perception of risk, 25–26
Performance of Health IT, 86, 106, 109–114
Performance testing, 238
Picture Archiving and Communication Systems (PACS), 6
Population at risk, 211–212
Post completion error, 73
Predictive behaviour, 75–76
Proactive controls, 178
Processing power, 113
Process variability, 135–137
Professional registration, 150, 151
Project change, 167–168
Project dependencies, 157
Project initiation document (PID), 157
Project management, 7, 130, 202

Q
Qualifications, 149, 151
Quality management, 47, 53–55
Quality policy, 55

R
Reactive controls, 178, 179, 284
Reasonably foreseeable misuse, 56
Recency bias, 215
Recognised Consensus Standard, 52
Recovery point objective (RPO), 108
Recovery time objective (RTO), 103
Redundancy, 82, 85, 101, 104–108, 191, 207, 224
Reference data, 63, 94, 96, 97, 194, 212, 294
Referential integrity, 89, 95, 96, 107
Reflective observation, 227
Regression testing, 237
Regulation, 13, 69, 77
Regulators, 17
Reliability, 81
Repeatability, 122
Residual risk, 43, 45, 52, 135, 178, 258, 268, 269, 271
Resilience, 62, 85, 101, 103, 104, 151, 224
Resiliency, 191
Resolution, 92–94
Responsibility agreements, 53
Retrospective data entry, 115
Retrospective safety analysis, 162–163
Ripple effect, 289
Risk, 23–24
  acceptability, 39–41, 43
  benefit analysis, 43–45
  versus benefits, 44–45
  compensation, 216
  equivalence, 37
  estimation, 205
  management file, 52, 137
  matrix, 32
  matrix categories, 32–34
  necessary, 25
  ownership, 43–44
  perception in Health IT, 26–27
  perception of, 25–26
  quantifying, 31–32
  risk matrix, 32–34, 36, 37, 186
  score, 31
  types of, 24–25
Risk evaluation
  qualitative, 32
  quantitative, 32–33
Risk priority number (RPN), 197, 198, 209
Root cause, 82
Root cause analysis, 82, 213
RPN. *See* Risk priority number (RPN)
RPO. *See* Recovery point objective (RPO)
RTO. *See* Recovery time objective (RTO)
Rule based performance, 67

S
Safety case
  analysis and evaluation, 268–269
  assumptions, 164–165, 267
  benefits, 171–172
  controls and validation, 269–270

Safety case (*cont.*)
in IEC 61508, 49
invalid, 136
maintaining, 287–297
managing change, 287–292
no go, 271–272
relationship to ALARP, 40
risk benefit analysis, 44
safety claims, 266
shortcomings, 172–173
staging, 272
structure, 265–271
writing style, 272–275
Safety culture, 65–66, 151, 186, 287
Safety integrity level, 49
Safety management file, 151
Safety management system (SMS), 292
accountability, 129–130
auditing, 131–133
documenting, 125–126
document management, 137
effectiveness, 133–135
governanace committee, 130–131
overview, 121–123
training on, 137–138
types of, 126–127
Safety requirements, 48, 63, 91, 158, 166, 189, 240, 241
Scalability, 113
Scale, 134, 211, 244, 278, 279, 282, 285
Schimmel, 4
Scope of analysis, 159
Screen size, 92
Security violation, 142
Semmelweis, 4
Senior management, 5, 53, 129, 130, 132, 271, 284
Sensitive data, 142
Service level agreements, 101, 191, 224
Service monitoring, 106–107, 224
Service outage
planned, 102
unplanned, 102
Severity categories, 31, 32
SFAIRP. *See* So far as is reasonably practicable (SFAIRP)
Single point of failure, 65, 104, 106, 152, 191, 200, 201, 217
Sittig and singh, 62
Situational awareness, 73
Skill based performance, 66
Skill mapping, 151
Skill mix, 151
Slips, 67–68, 72, 207, 214, 256
Slow system performance, 103, 109–114
SMS. *See* Safety management system (SMS)
SNOMED CT, 96
Socio-technical environment, 28, 61–65
So far as is reasonably practicable (SFAIRP), 40
Software as a service, 102
Software development lifecycle, 56
Software of unknown provenance, 57
SOP. *See* Standard operating procedures (SOP)
Sprint, 165–167, 236
SQEP. *See* Suitably qualified and experienced personnel (SQEP)
Staff harm, 144
Standard operating procedures (SOP), 231, 242, 259–261
Standards in Health IT, 47
Static testing, 234, 262
Storytelling, 195–197, 249, 280
Stress testing, 238
Structured What-If Technique (SWIFT), 183, 184, 262
breaking down the scope, 188
guidewords, 184–185
multidisciplinary approach, 185–186
top-down methodology, 186–187
Subjectivity, 274–275
Suitably qualified and experienced personnel (SQEP), 149
Survey, 134
SWIFT. *See* Structured What-If Technique (SWIFT)
Swiss cheese model, 61
System administrator, 87, 96, 151, 159, 191–192, 207, 255, 269, 296
System failure, 82
System testing, 162, 166, 237, 238, 269, 272

T

Test cases, 190, 239–243, 245
Test data, 239, 255, 257, 259
Testing software
background, 233–234
defect management, 243–245
design decisions, 242
documentation, 239–240
documenting in safety case, 245–246
environment, 238–239
inspection, 241–242
phases, 236–238
strategy, 240–241

types of, 234–236
Test plan, 186, 239–242, 245, 251, 255
Test report, 239
Test script, 235, 239
Thalidomide, 4
Think-aloud, 255
Time wasting, 110
To Err Is Human, 5
Top management, 65, 66, 127, 130
Traceability, 126, 167, 180, 219, 240, 243, 245, 249, 260, 290
Traceability matrix, 243, 245
Training courses, 226
Training materials, 70
Training plan, 151
Transactional demand, 111

U
Unique identifiers, 89
Unit testing, 237, 243
Upgrades, 86, 159
Usability, 55–56, 69–76, 89, 197, 214, 235, 236, 250–251, 278
engineering file, 56
interviews, 251
specification, 56
testing, 254–256
Use cases, 186, 190
Use error, 55, 61
User acceptance testing, 254
User-centred design, 70, 78, 214, 223, 225
User documentation, 70, 230
User interface design, 55, 69–76, 94
User interface evaluation, 250–251
User permissions, 87

V
Validation, 233
Verification, 233
Version control, 55, 96, 126, 132, 137, 159, 168
Violation, 68–69, 142, 144, 228, 229, 261
Virtualisation, 105, 113

W
Walkthrough, 234
Waterfall model, 165, 236
White box testing, 235
Workaround, 165, 206–209, 212, 282, 284
Workflow, 68
Workshop, 36, 192, 228, 262

Printed in the United States
By Bookmasters